T0289175

The Research Productivity of Scientists

How Gender, Organization Culture, and the Problem Choice Process Influence the Productivity of Scientists

Robert Leslie Fisher

University Press of America,® Inc.
Dallas · Lanham · Boulder · New York · Oxford

4501 Forbes Boulevard
Suite 200
Lanham, Maryland 20706
UPA Acquisitions Department (301) 459-3366

PO Box 317
Oxford
OX2 9RU, UK

Printed in the United States of America
British Library Cataloging in Publication Information Available

Library of Congress Control Number: 2004112551

ISBN 0-7618-3025-1 (paperback : alk. ppr.)
ISBN 0-7618-2942-3 (hardcover : alk. ppr.)

™ The paper used in this publication meets the minimum requirements of American National Standard for Information Sciences—Permanence of Paper for Printed Library Materials, ANSI Z39.48—1984

Dedicated to the Memory of My Parents
Sidney Fisher and Esther Schwartz Fisher

TABLE OF CONTENTS

LIST OF TABLES

PREFACE

This book is about the research productivity of scientists, and in particular, about the social factors that influence their productivity. The main idea driving the analysis is that productivity must be understood with reference to the research problem choice process in science, a process that is organizational and that involves decision making in conditions of uncertainty.

The departure point of this book is two observations. The first is a point made by Robert K. Merton, dean of American sociologists of science, who remarked that "scientists often do not know who else is engaged in similar work and this lack of information generates its own brand of pressures and anxieties." The second is an observation by a Dutch scholar, Sjerp Zeldenrust, who saw problem choices in science as resulting from organizational decisions made in ambiguous circumstances. He suggested that the ideas of the organization theorist James March and his collaborators, pioneers in the study of organization decision making in conditions of uncertainty, could be applied to the study of problem choices in science.

Social scientists have long understood that organizations have an important influence on their members. It is also well accepted that scientists are almost always organization members rather than solo practitioners. Yet, the two main perspectives on problem choice in science seem to ignore these simple truths. The rationalists emphasize that problem choice is an individual scientist decision made with reference to the views of professional peers. Any deviation from this way of deciding what to study, the rationalists believe, results in inferior scientific work since scientists know the significant problems of the discipline. An increasingly influential alternative view, the social constructivist perspective, emphasizes the chaotic and local nature of problem choices, although leading scholars championing this perspective, such as Karin Knorr-Cetina, acknowledge that multiple discoveries are possible.

In this book a point that will be emphasized is that scientists cannot be presumed beforehand to know what problems are important. It is usually in disciplines with strong paradigms that we find scientists asserting their prerogative to define what problems are worth studying. However, in fields without strong paradigms (e.g. most social sciences), scientists need and welcome guidance from nonscientists on what to study.

Several inferences follow from the assertion that scientists cannot be presumed beforehand to know what problems are important to study. One is that problem choices are likely to emerge out of interactions among scientists and laymen. In short, the decisions on what to study are likely to be organizational decisions and not individual scientist decisions.

A second inference is that there could be patterned differences in decisions about what to study that are determined by characteristics of the scientists. This point needs a bit of amplification. First, we must accept the idea that women scientists, having been socialized differently than men scientists, for example, bring different priorities to the discussions with laymen about what can and should be studied in fields without strong paradigms. Second, we must accept the idea that if multiple discoveries are possible then either organizations can have similar priorities or scientists can have similar priorities, perhaps based on gender or ethnicity. (In fields with strong paradigms, however, many scientists will have similar priorities, regardless of gender or ethnicity). As this study makes clear, there are gender differences in research priorities and in styles of research; and these gender differences influence the problems chosen. And these differences also bear on the research productivity of scientists.

Because the bulk of the book considers the question of what perspective is most likely to be a useful basis for a theory of problem choice in science, the presentation of the argument follows a somewhat tortuous path. For example, the discussion of the issue of organization influence on scientists is deferred until after a number of other issues are disposed of. Nevertheless, I hope that the main thesis is clear that discipline type, organization characteristics, and personal researcher styles and problem choice preferences are essential factors in productivity.

ACKNOWLEDGEMENTS

I could not have written this essay without the assistance of a great many people. It is a pleasant responsibility to record my debts to them here.

My interest in sociology was nurtured by teachers at City College of New York and Columbia University. I particularly wish to acknowledge Professors F. William Howton and Joseph Bensman at City College and Professors William J. Goode, Eugene Litwak, Theodore Caplow, Sigmund Diamond, Terrence Hopkins, Peter McHugh and Allan J. Barton at Columbia University. My abiding interest in organizational analysis was sparked by courses with Professors Caplow and Litwak. Professor William J. Goode taught a sometimes refractory young man how to think sociologically. I am deeply grateful to Professor Harriet Zuckerman for suggesting the research question and hypothesis that formed the basis for this research. A committee consisting of Professors Harrison C. White (chairman), Susan Lehmann, and Robert Freeland reviewed an early draft of this monograph and made numerous suggestions for improving it, especially in the literature review and methods sections.

Other people made important contributions to various phases of the research. Professor Mary Ruggie reviewed the questionnaire used in this study and made many suggestions that improved the final version. Dr. Joel L. Fisher, my brother, reviewed the proposal, the questionnaire, and the dissertation for clarity and made useful suggestions for improving all of them. I would also like to thank my friend Professor Carl M. Harris, lately of George Mason University, for reviewing the draft questionnaire. I gained important insights and deeper understanding of economists' perspectives from discussions with my friend Dr. Gregory M. Arluck of Yeshiva University (now with Citibank) and with Messrs. Olaf Hausgaard and Frederick Barney of the New York State Department of Public Service.

Dr. Joel Fisher, Professor Harris, my fiancée, now my wife, Shirley R. Anderson, Mr. Roger Benson, President of the NYS Public Employees Federation and many other helped me find suitable respondents for this research. I thank them for this critical assistance.

Special thanks also go to Joanne Ruppel first for careful data entry and preparation of a codebook of variables and then at the end of the project for her meticulous work formatting this monograph. And warm thanks to my typists Lorna Gillings, Deborah Carey, and Karen Thorpe. Lorna typed draft after draft of chapters of the first draft and kept her humor and composure while working under intense deadline pressure. Deborah took over the onerous ask of typing later versions of this manuscript when Ms. Gillings had to leave for family reasons. Karen worked sedulously on the Index.

I also gratefully acknowledge the assistance of Dr. Lawrence Lessner, a mathematical statistician who consulted on the multivariate analysis. Dr. Lessner, who brought his considerable experience in model building in biostatistics to bear on their sociological problem made an important contribution to the study.

I also hereby acknowledge the 107 respondents who agreed to be interviewed or to fill out lengthy questionnaires about the subject of this research. The blanket anonymity given them prevents my naming them here. However, this dissertation is in many respects their story and I hope it accurately reflects what they tried patiently to teach me.

Finally, whatever the merits of this study, I know all too well that the errors it still contains are too numerous to mention. I take full responsibility for those errors while hoping that the study's virtues are still evident to the reader.

CHAPTER ONE: INTRODUCTION TO THE STUDY

Numerous studies (Helmreich, Spence, Beane, Lucker, and Matthews [1980]; J. Cole and Zuckerman [1984]; Fox and Faver [1985]; Long [1992]; Sonnert and Holton [1996] cited in Valian (1998:262) show that there is a "gender gap" in scientific productivity. Valian (*loc. cit.*) says that "Taken together, studies measuring productivity in terms of quantity find that women typically publish 50 to 80 percent as much as men."

Sociologists have offered varied explanations of this "gender gap": "limited differences" in socialization (see J. Cole and Singer [1991] and J. Cole and Zuckerman [1984] cited in Valian [1998:263]) and structural problems such as differential access to resources for men and women. (see Feldt [1986], M.I.T. [1999], and Valian [1998:263]). Recent work on productivity (Valian [1998] converges with findings from studies in research problem choice, notably that of Zeldenrust (1990 Ph.D. Diss.; also see Jansen [1995]) in spotlighting the influence of "organization culture" on productivity. Valian (1998:263) remarks that women scientists are under represented on the faculties of elite institutions and over represented on the faculties of less prestigious institutions of higher learning. She asserts that "the culture at prestigious universities does not just help scholars be more productive than they would be at less elite institutions, it forces them to be." Zeldenrust (who never referred to "organization culture" in his study) or his part not only offered many examples of how different "driving forces" influenced problem choices.[1]

In fact, Zeldenrust went further, asserting that research problem choices resulted from organizational processes and not from individual scientist decisions. This argument about the nature of the problem choice was a leitmotif of his study.

If there is a difference between men and women in productivity, and if productivity and problem choice process are somehow connected, is it possible that there is a difference in problem choice preferences of men and women scientists. Those who maintain that problem choices are the rational activity of individual scientists might say "No." However, one clue that, in fact, there is a difference in men's and women's problem choice preferences is Valian's observation (1998:265) that several studies (Sonnert [1995]; Long [1992]; and Zuckerman [1987]) found that the quality of women's scientific publications exceeds that of men when measured by citation rate. She finds that "papers by male researchers were cited 60 to 98 percent as often as papers by female scholars."

I see this finding from several studies as evidence that women scientists have been pursuing a deliberate strategy of meticulously planning their research studies as a means of encouraging investment in their work. This strategy may be costly in terms of productivity. On the other hand, differentiating themselves from men in this way could confer a competitive advantage on the women in their struggle to get resources for their research work.

The above example illustrates a point I want to emphasize: there is a connection between the problem choice process and gender differences in productivity. The connection is both direct and indirect. Both, I believe, are influenced by organization characteristics. Furthermore, the problem choice process, in which organization wishes are mediated through scientists' problem-finding strategies, itself influences their productivity. (Although I concede socialization's role in productivity, it is exogenous in my study whose focus is productivity's relationship to the problem choice process and to organization variables).

I realize that my views about problem choice are unusual and require explanation. Therefore, I devote a great deal of effort to the task of assessing the relative merits of my perspective against the two principal perspectives on problem choice in science. This comparison of the merits of the three perspectives is necessary to bolster the plausibility of my perspective and prepare the groundwork for the multivariate analyses where I scrutinize the evidence for my perspective more intensively.

Some Preliminary Observations on the Nature of a Problem Choice

Is the selection of research problems in science a rational activity? For years, the conventional wisdom in sociology has been that it is, that if scientists are free to choose their own problems they will indeed select the best problems that they can. And it has been assumed that this was equally true of men and women scientists. Any differences in their choices of problems might be due to hindrances of some sort placed in the way of one gender, presumably women, that limited what they were allowed to do.

In my study, I found that women who were relatively free to choose their problems based their choices on different criteria than equally free men scientists. It seems plausible to presume that if the criteria used are different then the choices might also differ. Yet this flies in the face of the conventional wisdom and raises uncomfortable questions: Is one gender less rational than the other? Is problem choice not a rational activity after all?

I cannot give any sort of comprehensive answer to this question. However, my research shows that problem choice is not a rational activity. Neither is it random behavior. It is the result of an *organizational* process that is nonrandom and non-rational. Because it is an organizational process, we cannot presume that scientists are autonomous in problem choice (in the sense that scientists only need to account to others who like themselves are trained in science). In fact, because few sciences have powerful paradigms that can suggest problems we cannot presume that at the outset scientists know what is the significance of a particular question.

Perhaps scientists know only that someone is interested enough to pay them to solve a problem (if it is feasible to do this) or, to put it another way, if there is a technically feasible problem that might address the demand. This, however, is enough to form a linkage -- and that is fundamentally what a problem choice is in science.

How do scientists decide which problems they will seek to find a demand to study? Or alternatively, how do they decide which demands to address by suggesting a feasible problem? Were this a detective novel instead of a scholarly treatise and were I George Simenon (the French detective novelist), I might be tempted to sum up my answer to that question as follows, "Cherchez l'equipe de chercheurs!" (Direct your attention to the research team). By this I mean that the decision to select a research problem, - *i.e.*, to form a linkage - is not an individual decision but an organization decision. And I would add, "Look at the organizational environment in which the choice is made by the team. Because typically the team members are also members of a larger organization(s) and its values and needs influence their own."

This is not the whole story by any means; for example, the theories available in one's discipline from which to derive problems sometimes play a critical role. However, my highly condensed answer has the virtue of underlining the importance I place on the fact that the process of decision making in this organizational milieu is not "rational". This is implied in the original working title of this study, "Why Is Ms. Aristotle Rummaging In Mr. March's Garbage Can?" (The title was reluctantly dropped when my lovely wife pointed out that to the uninformed the title suggested that I had written a study of homelessness in ancient Greece)!

Those who are familiar with modern organization theory, will realize that the working title contains an important clue to what I am going to say in this study. Mr. March is James March, a renowned social science theorist, whose contributions include a "garbage can" model of universities, a kind of organization that he refers to as "organized anarchies." (See Cohen, M.D., March, J.G., and Olsen, J.P. 1972)

Ms. Aristotle is a somewhat humorous reference to the fact that my study concerns how the problems researched by scientists are chosen, whether these scientists are men or women. Why are scientists rummaging in Mr. March's "garbage can"? The answer, I maintain, is that the important business of selecting research problems for scientific investigation is better explained by March's "garbage can" model (as adapted by S. Zeldenrust [1990]) than by other current models (*e.g.,* the rationalist perspective identified with Zuckerman and Gieryn [1978] or constructivist models of Knorr-Cetina [1995], Latour [1987]), *etc.*)

The question of why the problems studied by men and women scientists are likely to differ is just one aspect of a broader concern of sociologists with building an empirical theory of problem choice in science. The importance of this aim is well understood within the discipline; as Zuckerman (1978:74) has trenchantly put it: ". . . sociologists have begun to adopt the self-exemplifying stance that problem choice must be a central problem in studies of scientific development." Numerous scholars, -- W. F. Ogburn, R. K. Merton, H. Zuckerman, K. Knorr-Cetina, to name but a few -- have addressed the issue. That this is a matter which has engaged the attention of so many sociologists must no doubt baffle the average man in the street. More than once when I was asked what I was researching and I replied "why do scientists choose the problems they select and ignore others my interlocutor would reply, "because they're paid to!" This commonsense answer is not wrong if we mean to confine the discussion to the *long* run and to the work of scientists in *general*. However, if we intend to answer the question in the *short* run, (where the answer really matters), and to apply it to actual researchers, then we will find the "commonsense" reply to be not at all adequate.

To this larger question of why scientists study the problems that they do I cannot give any sort of comprehensive answer in this study. Yet in trying to answer the more limited question of gender based differences that I posed at the outset, I discovered that I had to address matters that are central to the task of answering the bigger question. For example, I had to provide a definition of what a problem choice is, and this in turn necessitated my offering a novel conception of its nature. That conception, (based on Zeldenrust's work [1990,Ph.d. diss.]), is of a *linkage* among four independent or loosely coupled flows: (a) demands, (b) problems, (c) resources and (d) constraints. The model that I propose to explain gender based differences in problem research, *i.e.*, linkages, is not concerned with whether demand precedes problems. Neither is it concerned with where problems are obtained. It is simply concerned with factors that affect why a research team addresses particular demands and not others through the problems it studies.

In Chapter Six, I consider the implication of the organizational process of problem choice for researcher-productivity. Women scientists experience more interference from non-scientists in their research than men. Women seem to be more meticulous in their research planning than men and are less "opportunistic" than men in their selections. Can these kinds of differences be related to the "gender gap" in productivity that have been documented in numerous studies over many decades? The present study, based on a small sample, cannot answer this question in a definitive way. However, it seems that there is reason to believe these gender differences in research preferences and experiences as organization member are predictors of productivity.

Because of the inherent limitations of a small data set, I am only able to explore some of the relevant issues in the study of problem choice and its relationship to scientific productivity. Yet, in trying in my study to limn in some essential aspects of the "garbage can" model of problem choice, I believe I have pointed to promising lines of inquiry for future work. That is perhaps its most important contribution.

OVERVIEW

My study is divided into six chapters, including this one. In Chapter Two, I lay out the new perspective of problem choice I am proposing based on S. Zeldenrust's research (1990 Ph.D. diss.). Central to the new perspective is a definition of a choice radically different from the view implicit in the rationalist tradition of problem choice (*see* Zuckerman [1978]; *see also* Zeldenrust [1990 Ph.D. diss.]). This new perspective's intellectual roots in the "garbage can" model of March and his collaborators (1972, 1976, 1981) and in the "laboratory studies" tradition of sociology of science (*see* Knorr-Cetina [1995]) are described in detail to emphasize that the model sees a problem choice as *nonrational* and *nonrandom* behavior.

I also provide a review of the two principal current traditions of research of problem choice: the rationalist hypothesis and the social constructivist perspective. I point out that the former grew out of research in the history of science, sociological theory of professions, and utility theory. I also indicate that the social constructivist position, a sharply different view of scientist behavior, has its provenience in the phenomenological school of philosophy and in social anthropological research methods.

In Chapter Three, I describe the general approach of how I will assess which model is most promising as the basis for a theory of problem choice. My strategy is to distinguish five areas where the models should predict differently and then to specify hypotheses that should point to which model is better in that sphere. The five areas are: (1) Do scientists have autonomy in problem choice? (2) Are problems chosen for research derived from theory or locally originated? (3) Do scientists compete with other scientists to solve the same problems? (4) Is organization culture important in problem choice? (5) Is gender of the scientists, *i.e.,* the principal investigators, relevant to problem choice?

Since the various models sharply diverge on at least some of these five questions, the most promising model will be the one most consistent with the data from all of the examined spheres.

In Chapter Four, I discuss my methodological approach in more detail. I chose an unorthodox research method, albeit one with an impressive sociological pedigree. Numerous scholars with an anthropological bent or influenced by the ethnomethodological school (Harold Garfinkel, Barney Glaser, Julian Roth, David Sudnow to name just a few examples) have used this approach.

The best way to describe the study approach in the present study is to contrast it to the orthodox way of studying a problem.

The orthodox method to test a theory in social science is to do a survey study based on a large probability sample. This assures that sample statistics are representative of the parameter values of the larger population about which one wishes to draw inferences.

The present study is based on a non-probability sample (indeed, one that is deliberately biased in favor of highly experienced senior scientists). It is legitimate to ask if the results of this study are valid for a meaningful population. I argue that they are valid.

If the appropriate population were individual scientists, I would need a probability sample because of the heterogeneity of such a population. In my case, I maintain that the appropriate population is research teams. Teams are formal organizations in the sense that they have an individual identity, an exact roster of members, and an ability to act a corporate unit. Problem choices are team actions, taken perhaps after interactions with other members of some formal organizational entity. In this sense, scientists are not talking about themselves personally but about the teams of which they are principal investigators.

Regardless of what population the results of my inquiry may apply to I still had to confront knotty difficulties in gathering the data for this study. The first question I needed to address was how to construct a suitable population frame for this research. I had originally set out to study problem choice in multi-disciplinary teams of scientists. There is no readily available reference from which one can identify a population of scientists with problem selection in a multi-disciplinary research environment. It seemed to me that I would have enough challenge finding scientists who might meet my criteria for inclusion in the study.

In the framework of a probability sample study. answering the question of how many cases are enough is straightforward, an optimization problem in which one balances precision against cost. The greater the precision needed, the greater the number of cases required in the sample, and the greater the cost. Outside the probability sample context the question of how many cases are enough is pure guess work, more art than science. The advice of the scholar Anselm Strauss (*see* Strauss and Corbin [1990]) seemed helpful in this situation. Strauss advises the researcher to build a theoretical sample, making some educated guesses about which variables are likely to be important to understanding the phenomenon. I was able to obtain variation on three dimensions I believed were relevant to problem choice: (1) gender of my respondents; (2) discipline of my respondents; and (3) organization context in which they work.

The resulting sample of 107 cases, I am persuaded, is sufficiently large and heterogeneous that the data based on their responses provide an accurate picture of the process of problem choice in science. It is not a precisely accurate picture and certainly not a complete picture. Subsequent work may show that in some specific instances my views are incorrect in some way. However, a remarkable consensus across genders and disciplines emerged from my survey data that I believe represents a reasonably accurate understanding of how problems are chosen. My results, reported in Chapter Five, address the following questions:

(1) Are scientists free to choose their problems?
(2) Where do scientists obtain their problems?
(3) Is competition for priority among scientists an influence on problem choice?
(4) What role does the organization environment play in the problem choice process?
(5) What role does gender play in the problem choice process (and in scientist productivity)?

Since I argue that the various models have sharply divergent views on each of these five questions, empirical data on these questions should indicate which model is more promising as the basis for a theory of problem choice.

I present my findings on problem choice in Chapter Five. The principal finding of the study is the major importance of the organizations where the scientists work (or, more accurately, their organization of orientation) in the problem choice process.

One indicator of the organization's importance is found in the first noteworthy finding of my inquiry: women scientists report less freedom to choose their problems than men do. However, a substantial minority of all my respondents, both men and women, indicated that they are not free to choose problems for research. Other survey studies show even fewer scientists proportionately have freedom to select their problems than my own did.

The most striking evidence of organization effect on problem choice was my finding that scientists consider if there is available equipment that their employer wants to see utilized. A related finding is that the scientists look at whether using highly advanced research techniques (even if using the advanced techniques is not yielding more valuable knowledge than using elementary technology could).

These findings underscore a point emphasized by Karen Knorr-Cetina, a leading social constructivist scholar. She has asserted that besides scientists there are other participants who may have little or no background in science involved in the process of problem choice. It is easy to illustrate her observation. When the United States Government contracts for some study, not only the people interested in the study such as military leaders, but lawyers, budget analysts and personnelists directly or indirectly can influence the questions scientists on the research team actually address. These nonscientists accomplish this by setting limits on the funds available for the research, the qualifications of research and support personnel, and the time in which the study interval can occur. While the nonscientists influence problem choices in general, my findings that women report less freedom of choice than men suggests that the nonscientists may have a more decisive say in what problems are addressed by teams headed by women research than in the case of men scientists.

These particular findings may give some comfort to the rationalist scholars. However, some other findings of the study raise serious questions about how useful the rationalist hypothesis is as a way to describe problem choice in science. Before I offer them, I want to make an observation. The rationalists presume that scientists know the intellectual significance of problems before they commence their studies. Problems are said to be "shared" (Gieryn [1978] or "previously identified" (Zuckerman [1978]). This assumption of "shared" or "previously identified" problems is unjustified, according to my research. There are some fields where problems, for the most part, do not come from paradigms but rather from prior findings of the scientists or even from the scientist's realization of the need to fill a "theory gap." Medicine is a prime example of a field where prior findings rather than dominant paradigms may play a big role in problem finding.

Because of the lack of "shared" problems or "previously identified" problems in some fields, it cannot be presumed that scientists know the significance of questions ahead of time. Researchers in these circumstances may need to look to nonscientists for guidance on what is worth studying. They may even need to look to their own values or to theory gaps for inspiration. This point is illustrated by one of the most surprising findings of the study. Relatively autonomous women principal investigators based their research suggestions or preferences on different criteria than equally autonomous men.

My finding that scientists cannot be presumed to know ahead of time the significance of the problems they study foreshadows my next important finding. There is very little head-to-head competition in science. There are some obvious and not so obvious reasons for this. Head to head competition is costly. Furthermore, the rise of technology sharing agreements between rival firms which invest in each other's research in hopes of deriving ideas and products beneficial not only to the host company but also to its rival, also dampens the amount of head-to-head competition.

The rationalists emphasize the importance of competition for priority in understanding why scientists choose the most important problem they can. Solving important problems first confers prestige, fame, perhaps even wealth. However, my research does not support the rationalists' emphasis on competition for priority as an engine of scientific progress. Most scientists work in niches and do not see themselves in competition with others. They do not often acknowledge that competition is motivational for them.

An important caveat in these comments is that what is true among scientists is not applicable to the organizations where these scientists work. Problem choice, as I have repeatedly stated, is an organizational process. Competition with rival firms influences the actions of the company and no doubt also the research agenda of its scientists.

Up to now I have largely shown indirect evidence of the organization's impact on research problems scientists address. Many of my findings however directly illustrate the scope and depth of influence of the organization. Generally, these findings show the organization of orientation's influence on:

(1) the scientist's commitment to paradigms that she learned in her prior professional training. In extreme cases, she may come to see those paradigms as irrelevant or unimportant, if not actually incorrect.

(2) the scientist's willingness to embrace new paradigms important to the work that she does for the organization of orientation.

(3) the scientist's embracing new bases for assessing problem significance other than the professional criterion of theoretical importance of the problem.

Neither the rationalist nor social constructivist models correctly define the role of the organization of orientation in the problem choice process. The rationalist model largely ignores the role of the organization of orientation, preferring to focus on the characteristics of the problem itself in explaining why it has been chosen.

The social constructivists give "demand" a role in problem choice; however, they do not see any underlying regularities in the "chaotic" choice process that can be explained by specific characteristics of the organization or orientation. Only the "garbage can" model that I endorse assigns a key role to the organization's characteristics in helping to explain why problems with certain characteristics are chosen (while also acknowledging that the specific problem choice is unpredictable).

In Chapter Six, I return to the question that I began with in the Introduction: why do men and women scientists differ in their productivity? First, I tried to disentangle the impact of discipline type on productivity from that of gender since more men than women are drawn to disciplines of paradigms. Then I looked at whether this gender difference in productivity might be linked in some way to differences in problem choice preferences and differential experience with organization interference in the research experience. This should be especially noticeable in fields where there are no dominant paradigms from which to draw significant problems. This leaves room, I hypothesized, for problems to be suggested by the researcher's own values and organizational needs.

In addressing this hypothesis, I pursued two connected lines of inquiry. One line of inquiry was to look at whether differential experience with organization interference influenced gender differences in productivity. This inquiry was suggested by work of Valian (1998); Xie and Akin (1994); and Xie and Shauman (1998).

Secondly, I considered whether (a) differential experience with organization interference was related to different scientists preferences and behaviors regarding problem choices and (b) whether these preferences and behaviors, in turn, were predictive of productivity. Finally, I tried to assess the influence of organization characteristics and scientist preferences and behavior on productivity in a multi-variate analysis. Although the results are encouraging, it is not possible, given the inherent limitations of a small data set, to draw definitive conclusions. Still, I believe that future research building on the suggestions in this small study will demonstrate that the process of problem choice impacts men and women differently.

Women have experienced more interference in their professional work than men according to my research. They may find that to be taken seriously they have to be more meticulous, more respectable in their work from a methodological and theoretical standpoint than men need to. This colors their problem choices. Thus, I see the answer to the question of why there are gender differences in productivity as based on:

(1) Differential organization influence over problem choice of research teams headed by men and women and;

(2) Among men and women who are relatively autonomous men and women employ different criteria in selecting their research problems.

Now it is time to lay out my case for the "garbage can" perspective on problem choice, a theoretical point of view that I believe can point to a solution to some of the most intriguing questions that this study had to resolve.

CHAPTER TWO: MODELS OF PROBLEM CHOICE IN SCIENCE: PRINCIPAL HYPOTHESES AND HISTORICAL DEVELOPMENT

Far better an approximate answer to the right question, which is often vague, than an exact answer to the wrong questions, which can always be made precise.[1]

John Tukey
Statistician

Prefatory Remarks

While this study is entitled *The Research Productivity of Scientists*, it is really about two things. One is the role that gender differences in problem choice criteria and research styles play in research productivity. The other is the social determinants of those gender differences in problem choice criteria and research styles.

That men and women scientists choose different problems for research should surprise no one who has followed recent developments in the sociology of science. It is well known that there is a gender gap in productivity of scientists (Cole and Zuckerman [1984]; Xie and Akin [1994]; Xie and Shauman [1998]). Other research has shown that women scientists and men scientists are not treated equally in equipping their laboratories (Feldt 1986); and that women scientists are less likely to hear of opportunities informally and are thereby disadvantaged in doing research (Fox [1992:194-197]).

As Fox points out (Fox [1992]), research work occurs in organizational settings. And as Knorr-Cetina (1981) has remarked, before research is done, many people, including nonscientists, participate in the decision on what to study.

In short, given these diverse findings reported in the literature in the last twenty years on scientist productivity, the possibility that someone would find that men and women scientists prefer different problems for research seems almost inevitable. Yet, until now, no one has even suggested, much less demonstrated, that such gender based differences in problem choice occur. Obviously, neither has anyone offered an explanation for this difference in problem choice criteria.

Equally obviously, it is important to both demonstrate that there are gender based differences in problem choice, and to explain them. I maintain that these problem choice differences are relevant to productivity differences between men and women researchers—a point I develop further in Chapter Six. I also maintain that by explaining these problem choice differences in terms of the "garbage can" model of organization behavior, I am not only closing a gap in knowledge about problem choice but also linking a major research focus of the sociology of science to the powerful models in organization studies, in particular the "garbage can" models of March and his collaborators.

The task of explaining gender differences in problem choice that I have undertaken is a response to Harriet Zuckerman's (1978) question,". . . how do scientist choose from the pool of identified problems?" In order to answer her rhetorical question, I needed to use a novel conception of choice that S. Zeldenrust (1990, Ph. D. diss.) developed, based on theoretical work by James March *et al.* (1972, 1981). Zeldenrust essentially says that when scientists find a demand and also identify a problem that meets the demand, by that linkage of demand and problem, they have made at least a preliminary choice.[2]

In arguing that the choice is a *linkage* between a demand and a problem, I adopt a radically sociological view quite different from the prevailing "rationalist" perspective of problem choice. My view is radically sociological because I emphasize that the choice is *culturally* determined; it is not only the characteristics of the problem that are influential in the choice, but also the characteristics of the *entire choice* situation including, for example, the gender of the principal researcher and organization culture of the setting in which the problem choice is made.

The sociological perspective I offer in answer to Zuckerman's question is not a complete theory of problem choice. Given the current state of technical development in sociology, I do not think a complete theory is possible right now. I only hope to set right the theoretical underpinnings, thereby permitting a correct answer to emerge eventually from subsequent research.

Section One: An Organization Studies Perspective to How Scientists' Research Problems are Chosen

I have already stated that a choice in my view is a linkage. More precisely, I conceptualize a problem choice, (following Zeldenrust [1990 Ph.D. diss.]) as a linkage among four independent (or at most loosely coupled) flows: (a) demands, (b) problems, (c) resources and (d) constraints.[3] The choice is a result of organizational, not individual scientist decisions, reached after a process that entails interaction within and between the research organizations and their environments.[4]

Because problem choice is an organizational decision, based on imperfect information and made in a climate of (1) conflict between preferences and (2) changing preferences over time, it is *not* a rational decision as this term is understood in utility theory (Fishburn [1977]).

The perspective I advocate is not easy for sociologists to accept. For example, Gouldner (1970:55) observes:

> It cannot be stressed too strongly that in everyday practice the sociologist believes himself capable of making *hundreds of purely rational decisions -- the choice of research problems*, sites, questions, formulations, statistical tests, or sampling methods. He thinks of these as free technical decisions and of himself as acting in conformity with technical standards, rather than as a creature molded by social structure and culture [emphasis supplied].

The dominant sociological paradigm of problem choice conforms closely to Gouldner's depiction of how sociologists perceive their professional work, *i.e.*, "research is seen as an individual, autonomous and *rational* activity in the context of stratified science fields." (Zeldenrust [1990:3], emphasis supplied).

In contrast to this dominant view, Zeldenrust offers a view of a problem choice as non-rational and nonrandom. I will sketch this new conception here closely following Zeldenrust's development.

Zeldenrust (1990:30) carefully distinguishes between research as a "problem solving process" on one hand and a "choice process" on the other. The former refers to the relationship between "research problems and subsequent outcomes' and belongs to "the psychological study of scientific creativity." The latter refers to "the context in which problem solving takes place and which shapes its orientation." He adds that these "choices direct problem solving activity."

The second element is the choice process, *i.e.*, "the context in which problem solving takes place and which shapes its orientation." These "choices direct problem solving activity (1990:3).

It would be a hopeless task to explain the new perspective of problem choice without first pointing out, as Zeldenrust has done(1990:4), that models of problem choice can be formulated at "different levels of analysis."

The lowest level concerns the negotiation of problems and "demands." This level, explored by Fujimura (1987) and Knorr (1974), also is suitable for the network analysis concepts of theorists such as White (1992:esp. Chap. 1 through 4). A second level of analysis, according to Zeldenrust (1990:4), is the "choice arena."[5] Zeldenrust describes this as analysis of "the research group as a 'local heuristic organization.'" He says further that in this arena ". . . various demands and research problems and . . . elements such as resources and constraints enter, leave or circulate. In certain instances, a demand and a problem may team up into a *linkage*: this linkage is in fact a choice (emphasis original)."

The third level of analysis is "the degree of *control* that heuristic organizations have over sets of research problems, demands, and other elements such as resources." The organizations may have control over problems but not demands, for example, or over demands and not problems, *etc*. The importance of this control is in the probability of a linkage being formed between the flows. Note that for Zeldenrust, the control is exercised by a research group. However, it seems that, in actual practice and logic, an individual researcher can decide on a research problem just as in White's [1992:53] example of the chief executive officer (C.E.O.) either an executive committee or a single president (as C.E.O.) can decide on a course of action for the company.

Having specified the different levels of analysis that are possible, Zeldenrust then turns to the development of his model. The first significant point about his model that I wish to make is that Zeldenrust borrows the concept of "streams" or "flows" of independent processes from March *et al.* work. In the latter's work (*see* Cohen, March *et al.*, 1972; Cohen, March and Olsen, 1976), the "garbage can" model originally consisted of "four streams or flows of elements over time." In their work, while noting that "in a particular choice arena there can be several relatively independent flows," Cohen, March and Olsen (1972) focused on "the flows of problems, solutions, choice opportunities, and participants."

Zeldenrust modifies this set of processes for his model. However, as in Cohen, March et al. (1972), a key assumption in this "garbage can" model is that the streams or flows are "independent or loosely coupled." From this it follows that streams do not have temporal ordering: for instance, in the Zeldenrust model, the research problems stream does not necessarily precede the problem solutions stream. Furthermore, as Zeldenrust points out (1990:15), a numerical imbalance between the problems and solutions can easily arise because any of the following are possible. (1) "problems may emerge, disappear (without being solved) or become modified over time," (2) "solutions may enter and again leave without being adopted, and adopted solutions may not even solve a problem" or (3) both changes in problems and solutions may occur simultaneously. In this situation, a choice of some action (not necessarily a research choice) is, as noted by Cohen, March, and Olsen (1976:27; quoted in Zeldenrust *loc. cit.*) "…a somewhat fortuitous confluence. It is a highly contextual event, depending on the pattern of flows in the several streams.

It is important to say here that Zeldenrust's model, while clearly influenced by the work of March and his collaborators, is not a carbon copy of it. Zeldenrust points out that Cohen, March, *et al.*, concept of choice "as a fortuitous confluence of elements" resembles his own view of research choice as a linkage in a key way; however, Zeldenrust adds that the linkage serves as a context which orients further (problem-solving) activity.

Next, Zeldenrust proposes some modifications of the "garbage can" model that I endorse. While the "garbage can" model's fundamental assumption is that the flows are independent or loosely coupled, Zeldenrust points out that "there may for instance be a systematic *coupling* between the flows in investigated cases." The typical case is that of "demand" coupled to "resources" and, thus, "one cannot enter into a linkage without the other." Second, Zeldenrust points out that "one or more flows may be *stable* or consistent. On occasion, for instance, the flow of problems may consist of just one, stable, consistent problem."

Zeldenrust makes an additional suggestion regarding a need for a process perspective. He points out that a Markov chainlike process sometimes is found in research decisions because some choices "constrain or enable subsequent choices." Thus, "there may be more stability, logic or consistency" in the research decisions than suggested by the model

March and his collaborators proposed for university administrative decision making. Occasionally, "band wagoning" (*see* Fujimura [1988]) or perhaps another process "may produce major ruptures in chains of choices."

Foremost among the assumptions underlying the model is that there is ambiguity rather than perfect information.

Four kinds of ambiguity characterize the choice process,[6] according to Zeldenrust, who leans heavily on work of March and Olsen (1976, Chap. 1):

* intention
* understanding
* organization
* history

Intentions, Zeldenrust says, "are not necessarily clear, consistent or stable, and may follow rather than determine the choices that occur."

Ambiguity of understanding "includes difficulties in interpreting the organizational environment;" *e.g.*, this ambiguity results from "lack of understanding of 'technologies' (not just hardware, but the organization's internal processes of converting inputs into outputs) and lack of information generally."

The third type of ambiguity, *i.e.*, ambiguity of organization refers to "varying attention given by participants (who are likely to have different preferences) to certain issues."

Finally, the fourth type, ambiguity of history, refers to the fact that "it is unclear what happened, how, and why."

Zeldenrust then shows in further detail how ambiguity is relevant to his model. For example, the "demand" stream, which corresponds to the concept of "problems" in the March *et al.* "garbage can" model, has four dimensions:

- urgency
- specificity
- level, and;
- multiplicity

each of which have "a degree of uncertainty (or lack of clarity) about their precise nature." (Emphasis original).

However, uncertainty is not the only way in which ambiguity is manifested. Zeldenrust also points out that "there may be shifts and conflicts" in demand which contribute to the ambiguity as well.

Ambiguity, in the form of uncertainty, shifts, and conflicts also affects all the other flows of the Zeldenrust model: resources (equivalent to "attention" in March *et al.* model); constraints (for which no equivalent concept if found in the March *et al.* model); and problems (equivalent to solutions in the March *et al.* model).[7] He says, (1990:23),

> Each of these flows may be uncertain (e.g. what exactly is . . . the "robustness" of a constraint or the "convertibility" of a resource?), they may shift (e.g. more demands or resources may enter, or demand may leave, or a constraint may change its value), and there may be conflict within or between flows (demands may conflict or compete, a demand and a constraint may conflict. (Emphasis in original.)

The central role that Zeldenrust assigns to ambiguity fundamentally differentiates the model, a kind of prospective theory of choice model (*see* Adelman [1997]; Tversky [1981]), from the rationalist model of problem choice suggested by Zuckerman (1978) and Gieryn (1978). The latter model, a utility theory model, *assumes perfect information about the significance of problems.*

In contrast to that rationalist model, in which important problems are unambiguous, and known to all; and where demand is assumed to be present for solution of these important problems and, thus, is likewise unambiguous; and where researchers are motivated to work on these significant problems in order to gain priority for their work,[8] the Zeldenrust model allows for ambiguity with regard to any or all of the following: importance of problems, disciplinarians' prior knowledge of problems, and demand for their solution.

There are important implications to the great emphasis on the role of ambiguity in the Zeldenrust model. For example, demand plays a central, explicit role in the model, whereas it plays hardly any role (and is not explicitly even acknowledged as relevant except in reference to military needs) in Zuckerman's model.

In Zeldenrust's "garbage can" model *demand* guides the researchers in searching for problems in the absence of information about the importance of problems, indeed, in the absence of information that a doable problem even exists to address the demand. Therefore, for sociologists interested in building a model of problem choice based on the "garbage can" model, contextual factors influencing changes in *demand* would be salient in their search for an explanation of problem choices.

In contrast, Zuckerman, who argues that problems are almost always derived from theories would lean in the direction of detailed studies of the development of and changes in paradigms for clues to scientist problem choices. (In actual studies, of course, since she and her disciples acknowledge that problem choice is affected by external factors (*e.g.*, military need), there would be at least some consideration given to clearly relevant external influences. Thus, in reality, the difference is a matter of degree, a matter of emphasis, rather than a sharply defined cleavage). It is evident that the two models diverge in their assessments of the fruitfulness of different lines of inquiry to developing the theory of research problem choice. Before I discuss this issue further, however, I want to turn to a review of the rationalist model, its history and the main lines of criticism leveled at the rationalist model. I shall do this in the next Section.

Overview of Section Two

In this Section, I set forth the rationalist perspective's principal hypothesis and key assumptions. I then contrast it with the views of its critics in the resource dependence camp and the constructivist camp. Finally, I suggest that the "garbage can" model, presented in Chapter One, Section One, is a perspective that addresses the weaknesses of the rationalist perspective while recognizing that the latter gives approximately correct answers in certain situations.

1. The Specter of Karl Marx

Why do scientists choose some problems and ignore others? Sociological concern with this question goes back to at least the 1930s in the United States. As the close of the twentieth century approaches sociological research generally employs one of three perspectives:

(1) the rationalist perspective of H. Zuckerman (1978);
(2) the "resource dependence" perspective represented by Pfeffer and Salancik (1978) (see Shenhav (1985 Ph.D. diss.); and
(3) The constructivist perspective identified with such writers as Knorr-Cetina (1982;1995).

All of these perspectives are related in some way to Karl Marx's views on science and scientific ideas.

Marx saw science and other spheres of society as intimately connected and mutually dependent upon each other. Thus, in *The German*

Ideology (1964; quoted in Shenhav [1985:6]) he wrote, "Where would natural science be without industry and commerce? Even this 'pure' natural science is provided with an aim, as with its materials only through trade and industry, through the . . . activity of men."

However, Marx also recognized the enormous influence of science on society; he wrote, for example, (1964; originally 1844:[83]) ". . . the natural sciences have penetrated all the more *practically* into human life, through their transformation of industry". He expressed his fervent conviction that "they [sciences] have prepared the emancipation of humanity, even though their immediate effect may have been to accentuate the dehumanizing of man (1964:73)."

Marx wanted to see the creation of a single human science and thus he seems to have disdained efforts by scientists to appropriate specific problem areas as their own while ignoring other areas of human activity. For example he upbraided the infant science of psychology, saying (*loc. cit.*).

> No *psychology* for which this book, *i.e.* the most tangible and accessible part of history, remains closed can become a genuine science with a real content. What is to be thought of a science which remains aloof from this enormous field of human work, of a science which does not recognize its own inadequacy, so long as such a great wealth of human activity means nothing to it, except perhaps what can be expressed in one word - 'need' or 'common need'?

One further aspect of Marx's thinking with regard to science is pertinent here. Marx emphasized the social nature of science; its ideas, he maintains, are a product of the real social conditions of the society in which the scientific work is done. He says (1964:77):

> Even when I carry out scientific work, an activity which I can seldom conduct in direct association with other men - I perform a *social,* because human, act. It is not only the material of my activity - like the language itself which the thinker uses - which is given to me as a social product. My *own* existence *is* a social activity. For this reason, what I myself produce, I produce for society and with the consciousness of acting as a social being.

2. The Rationalist Perspective:

Principal Hypothesis and Assumptions

A. Assumption of Scientist Autonomy

A key *explicit* assumption in the rationalist model is that scientists are free to choose their problems. "Free" in this instance means professionally autonomous, *i.e.* free of any constraints other than their responsibility as scientists to do good scientific work. Only in the case of autonomous scientists does the hypothesis predict that scientists choose the best problems they can.

The rationalists' assumption of autonomy is, in a sense, a reaction to Marx's conviction that science and other spheres of society should be closely linked; on one hand, the rationalists say autonomy is necessary for science to develop and, on the other, they also claim it is necessary to permit this autonomy to ensure that scientists work on problems to improve society.

Shenhav (1985:6), in a review of sociological perspectives on scientific development notes that:

> The underlying assumption of the structural-functional (*i.e.* rationalist) perspective in the sociology of science is that absolute autonomy of professional science is a necessary condition for the development of knowledge. According to that view, the invisible hand of theoretical needs regulates scientific progress (Polanyi [1951]; Sutton [1984]) and therefore institutional autonomy and internal normative structure are necessities for fruitful research.[9]

There is evidence that this assumption of autonomy in problem choice, *i.e.* control by scientist themselves of their work, colors Zuckerman's thinking in her statement of the rationalist perspective (1978); consider for example her observation (1978:82) that in studies by Edge and Mulkay (1970) and Lederberg and Zuckerman (1974) ". . . two criteria were most frequently used in selecting from arrays of previously identified problems: (1) the assessed scientific importance of a problem (which of course reflects theoretical commitments) and (2) the feasibility of arriving at solutions . . . ."

These criteria were obviously the sort that only *technically competent* scientists could have applied and, thus, their use would exemplify scientist control of problem choice.[10]

Despite the importance that the rationalist perspective attaches to the idea that scientists must have control of their problems, I found only one empirical study (Debaekere and Rappa [1994]) suggesting that there might be validity to this supposition.[11]

B. Assumption - The Nature of a Problem Choice

The rationalists conceive of problem choice as a selection from a finite list of describable problems. (Note Zuckerman's comment [1978:82], "two criteria were most frequently used in selecting from *arrays of previously identified problems*" [emphasis added]).

A second, implicit, assumption, however, is also contained in the hypothesis: scientists *know* what are the most important and challenging problems among the array of problems available for them to solve[12] This assumption that scientists can at least roughly classify problems into more or less significant problems leads to one of the most interesting predictions of the rationalist hypothesis: that intense competition for priority will occur as scientists try to solve the most important problems they can ,

In contrast, the "garbage can" model makes no prediction of intense competition for priority among scientists. This is because the "garbage can " model argues that "ambiguity" characterizes assessments of importance of any problem. There may be no real appreciation of how significant a problem is before it is solved. And even after it is solved, its importance may not be understood for a very long time.[13]

Zuckerman does not say how the scientist makes her choice of problems from among all "previously identified problems" endorsed by renowned researchers. One clue is her comment [1978:82] that "theoretical commitments" are central to this decision. However, this does not rule out other commitments not of a theoretical nature. For instance, the scientist may be committed to earning a lot of money in a hurry. (This may not necessarily reflect personal avarice; it could be that an especially lucrative project is chosen because it will help finance work on less financially rewarding but scientifically interesting problems.)

Zuckerman does not consider the process by which the choice is made. Is it a sudden decision or is it reached after a careful investigation? She says the problem must be amenable to investigation but perhaps the decision is made at least tentatively before this is clear or before strong evidence that it is *not* amenable to investigation, *i.e.* not technically feasible, is available.

In my early planning of what eventually became the present study I attempted to "fill in the blanks" regarding the process of problem choice, presuming that scientists behaved as the rationalist model postulates. I conjectured at the outset that scientists were rational decision makers operating with imperfect information. They wished to reduce uncertainty before committing substantial resources to a problem. Thus they would appraise it against various criteria:

* technical feasibility
* political-legal feasibility
* organizational feasibility

By posting that "overall utility" across the three types of feasibility was to be maximized, I relaxed the requirement that the project be technically feasible before it is undertaken. I shall refer to this version of the rationalist hypothesis from here on as the Lederberg- Zuckerman-Fisher hypothesis (or Model "L-Z-F").

C. The Rationalist Hypothesis

The general hypothesis of the rationalist model (in the Lederberg-Zuckerman-Fisher version) is that the chosen problem is derived from the identified stock of problems and has greater overall utility at the time the choice is made. Utility is assessed by the scientist against three spheres: (a) technical (is the problem apparently technically feasible?); (b) political-legal (is the research on this problem likely to engender unacceptable difficulties of a political and/or legal nature?); (c) organizational (is work on this problem likely to jeopardize other critical agendas of the research agency? Is there adequate commitment from the organization to support work on this problem?)

In Chapter Two I state the operationalized version of this general model which is in essence a simple utilitarian model melded with a conception that scientific problems are derived from theory.

What makes problem choice interesting is the *context* in which the problem choice is made: scientists are engaged in competition for "priority" in solving important problems. When the nation's scientists are choosing and solving important questions the country's scientific eminence is enhanced. And national leaders understand that scientific eminence is critical to national economic prosperity and political power in the world arena. Therefore, identifying and solving important scientific questions is a matter of some national importance. It is these considerations that make it important to understand factors that scientists take into account in appraising a problem before they commit resources to solving it.

3. History of Rationalist Research

Did the rationalist perspective spring full blown from Zuckerman's head one fine day in the 1970s perhaps while she was strolling the grounds of Columbia University's bucolic campus on Manhattan's upper West Side? Obviously not, though, I may have unwittingly fostered that impression in my discussion to this point.

The origins of the rationalist perspective are in sociological and historical scholarship beginning in the 1930s when sociologists began to look at the development of science in society. R. K. Merton (1935; 1938), S. C. Gifillan (1935a; 1935b), and W. F. Ogburn (1937) led the way. Even earlier according to Barber (1952) B. J. Stern (1927) had written a book, *Social Factors in Medical Progress*.

It is convenient to distinguish two periods in the development of the rationalist perspective: (a) a "pioneer" period from the 1930s to about 1962 and (b) the period since 1962.

From the standpoint of the development of the rationalist perspective, the most notable event in the pioneer period was R. K. Merton's publishing his influential dissertation, *Science, Technology, and Society in Seventeenth Century England* (1938). However, a less well-known contribution by B. Hessen (1931) as well as other Russian social scientists published in *Science at the Crossroads* (1931) are significant because they influenced the thinking of Merton, and J. D. Bernal [1939] at that time (*see* Barber [1952] and Shenhav [1985]).

All of the above-named scholars--Stern (1927), Merton (1935;1938) and Hessen (1931) assigned a prominent role to external factors in scientific development (*i.e.*, those outside science as an "institution"[14]) suggesting that from the outset a concern with external factors was characteristic of research in the rationalist tradition.

Despite this promising start for the specialty, research in the sociology of science languished for many years afterward; as late as 1952, Merton (1973:210-220; originally 1952) lamented that it was a "neglected" specialty. Then, quite suddenly, interest in the specialty bloomed just a decade later.

Zuckerman (1988:512) notes that:

> . . . two theoretical developments in the emerging specialty (of the sociology of science) sharpened its cognitive focus: the publication in 1957 of Merton's 'Priorities in Scientific Discovery' and in 1962 of Kuhn's *Structure of Scientific Revolutions.*

> Each set out new perspectives on the social organization of scientific inquiry and its patterns of growth, and each lead to major lines of inquiry.

What aroused so much interest in Kuhn's slim volume, *The Structure of Scientific Revolution*, was his argument that the history of sciences is a history of periods characterized by research based on a dominant paradigm. The change from one dominant paradigm in a discipline to another historically has been revolutionary at times--a sudden sharp change rather than an evolution from a long dominant paradigm to a new paradigm.

Perhaps most provocative of all Kuhn's points from the standpoint of the present paper was Kuhn's argument that paradigmatic change represented a change in how the same phenomenon were perceived. During the period when the old paradigm was being challenged by the new one, the adherents of the two paradigms would 'talk past each other' (Zuckerman, 1978).

Kuhn's views were not met with universal acclaim within the discipline of history of science and in other fields, notably sociology (Karl Popper, among others, was dismissive of Kuhn's work) and they remain controversial to this day.[15] But publication of Kuhn's work, seen from the vantage point of over thirty years later, can be considered a watershed in the sociology of science. After his monograph appeared, the pace of publication in the sociology of science seems to have increased. No doubt the fact that Robert K. Merton, the doyen of sociologists of science, shared some of the same views as Kuhn (see Zuckerman, 1988) was a factor in winning support for Kuhn's view. Papers on social structure of science and other issues suggested by the seminal work of Kuhn or Merton began to appear in the once largely neglected sociology of science. Zuckerman (1978:86) taking note of the accelerating pace of research in the sociology of science since the 1960s cautioned that ". . . systematic investigation of the cognitive and social sources of decisions that ultimately make for cognitive change in science was still at an early stage."

In Zuckerman's opinion, a logical step prior to the study of cognitive change in science would be research illuminating the bases of theory choice and problem choice in science. She says, (*loc. cit.*) however, that "sociologists have just begun to take steps toward describing patterns of theory choice and problem choice for individuals and for aggregates of scientists."

Kuhn's thesis, in my opinion, had one mischievous effect. While it stimulated research, much of that work focused exclusively on what has come to be called the "internal" factors in science that affect problem choice. Other work in this "internalist" vein include Koestler, *et al.*, (1982) which stresses the characteristics of a theory in determining physicists' enthusiasm for it and thus its general acceptance as a model for the phenomena the theory explains. And Fujimura's work (Fujimura, 1986) on the reasons for the acceptance of the oncogene theory in cancer research also points to the critical role of the characteristics of an idea in its acceptance by scientists. Zuckerman (1978:74) also lists some earlier papers.

Lederberg and Zuckerman's own work (1974 unpubl. MS) I would classify as in the "internalist" tradition. For example, in their hypothesis that problem choice is influenced by feasibility and assessed scientific importance, the concept of "feasibility" seems primarily to refer to the *technical* dimension of this idea and ignores other possibly relevant aspects of feasibility. And their variable of "scientific importance" placing intellectual criteria of science paramount in decisions about which feasible problem to research also seems to be an "internal factor."

The quantity and quality of research in the 1960s and 1970s emphasizing factors internal to science in problem choice probably colored Zuckerman's thinking about the relative *primacy* of internal and external influences. Despite Zuckerman's (1978:82) explicit recognition that external factors could influence problem choice, she suggested such a limited, and, to my mind, roundabout avenue by which external factors can influence problem choices that it would be easy to wonder about the depth of her conviction regarding a role for external factors. Thus, she writes (1978:82) that in studies by Edge and Mulkay (1976) and by Lederberg and Zuckerman (1974) "two criteria were most frequently used in selecting from arrays of previously identified problems." These were ". . . the assessed scientific importance of a problem (which, of course, reflects theoretical commitments) and the feasibility of arriving at solutions."

She then adds (*loc. cit.*):

> . . . explicit and tacit cognitive criteria of problem finding, problem relevance, and *problem significance* could in principle be related to the *social* as well as the cognitive structure of the sciences under examination. Along with this intricate research agenda is the related question of extra scientific influences upon problem choice. (emphasis supplied).

Zuckerman does not elucidate how scientific importance might be affected by extra scientific factors of any sort much less how the social structure might affect "importance." She is also silent on the possibility that social structure and other "extra scientific" factors can affect feasibility. Perhaps given her provisional definition of feasibility - amenability to scientific investigation - it would be hard to imagine a role for extra scientific factors.

I share Zuckerman's conviction that extra scientific factors *are* important in problem choice. But I question her suggestion that external factors indirectly affect problem choice through their influence on "problem importance." I believe that technically it would be difficult to demonstrate that assessed scientific importance is causal for problem choice. The difficulty arises partly because the opposite may be true: problem choice is possibly causal for "assessed scientific importance."[16]

Though [as I indicated before] the majority of studies of problem choice in the rationalist tradition were "internalist" in the post Kuhnian era (*see, e.g.*, Weinberg [1977] on the importance of whether a question even makes sense; Koester, White and Sullivan [1982] on physicist concerns with the renormalizability of a model of subatomic physics; Fujimura [1986] on the reasons why the oncogene model became dominant in oncological studies), there were a few that pointed to a role for external factors. Historical research by British scholars in this period, reviewed by McLeod (1977), exemplifies this. McLeod says, (1977:63):

> . . . Schofield's . . . study of the Lunar Society (1963), Shapin's study of the Royal Society of Edinburgh (1971) and of provincial scientific institutions in England (1972) and Morris Berman's analysis of the Royal Institution (1972) together reveal close interpenetration of political sympathies, personal ideas and philosophical assumptions in scientific research, a phenomenon further supported by very tangible interests of class and social status. One now sees the vicissitudes of the early Royal Institution in a totally new light, knowing that the early managers and proprietors considered investment in science to be a means not only of exploiting natural resources, but also of improving rural incomes, diminishing the burden of the poor and containing the possibilities of a popular revolution which could be sparked off by invasion from France!

A monograph by Gaston (1973) can also be mentioned here. Gaston studied the structure of the British high energy physics research industry and found that each research center was constrained by Government edict to address only a narrow range of problems in order to avoid wasteful competition. Gaston further states (1973:177 fn.) that the organization of American high energy physics research was beginning to move in a similar direction to the structure of U.K. research in this discipline. But this is not the only example of how non-technical factors may intervene in problem choice. Loren Graham (1987, esp. Chap. 2) shows, for example, how organizational cultures could also constrain problem choice in genetics research. According to Graham (1987:63), during the two decades when Trofim Lysenko was the political boss of Soviet biology, Soviet genetics researchers had two basic options. Under Option A they could try to do genetics research and publish their research in biology journals controlled by Lysenko and his allies. However, this was only possible if they repudiated Mendelian theory to which Lysenko was opposed. (Genetic research under this constraint was *not* impossible if the researchers were content to do applied agronomic work in such useful areas as livestock or seed improvement. Lysenko, though a mediocre scientist at best, endorsed ideas generally current among agronomists of his era). Or, under Option B, they could seek sponsorship for Mendelian influenced research from academic physicists or mathematicians, and could publish such work in cybernetics journals (which few Soviet biologists of that era could read since for the most part the mathematical background needed to follow developments in this field was greater than they possessed).

It is clear from Graham's discussion that the organization culture of particular Soviet research institutes, and not overall political arrangements, as in the U.K. example, was the relevant contingency for Soviet genetics researchers of the 1940s and 1950s.[17] This is because Mendelian genetics research was *not* stifled across the board (as would be expected if a political culture variable were operating). It is simply that while Lysenkoism was enjoying its heyday, scholars doing research based on Mendelian ideas found that they had to direct their results to scientific journals read only by a small scientific elite that appreciated these ideas.

4. Alternative Models of Problem Choice

A. The Resource Dependence Model

Not all the research done on external factors in problem choice can be called consistent with the rationalist perspective. Some ideas challenge the assumption of the rationalist perspective that research problem choice is in the hands of scientists. This strand of research, in a sense, updates Karl Marx's argument about the close interconnection between science and other spheres of human activity. (Marx enthused [1964; originally 1844] about this believing that in the future, after the Communist Revolution, the close interpenetration would only lead to mutual benefit [*see supra* p. 3]. Modern research, devoid of this utopian strain in Marx's thought, does, however, illustrate the close interpenetration of science and society and its effects on problem choice).

Shenhav (1985, Ph.D. diss.) reviewed many of these studies. He found that published analyses of the "impact of external material resources on scientific development" did not show uniform results; nevertheless, he concluded, they "indicate that external influences do exist; and that science and society are indeed interconnected." (Shenhav [1985:11]; see also Narin [1976]; Frame [1982]; and Useem [1976a.1976b]; for contrary evidence, see Thomas [1982] as well as Menard [1971] and Stolti-Heiskanen [1979] referenced in Shenhav [*loc. cit.*]).

Shenhav's findings, (1985, especially his Chapter 6) are most germane to the present study, and I will describe them in some detail. Shenhav investigated two hypotheses of "problem choice processes": Scientists in externally funded research settings are (a) less likely to be influenced by the scientific literature and (b) more likely to be influenced by external suggestions, in problem choice processes.

For his empirical data, Shenhav had results of a survey of Israeli scientists performed in 1976-1977 by A. L. Goldberg. Only the results on the 330 researchers in academic institutions are reported. (Shenhav provides no procedural information on this survey *e.g.*, [a] did the study rely on a random sample? [b] was there any sample attrition? or [c] was the instrument validated?).

Shenhav found for his study population of Israeli academicians that scientists are less likely to "feel autonomy in problem choice" the greater their dependence on external funds. The relationship is *monotonically* decreasing from 53% claiming to feel free to choose when no external funding is provided to 28% when all funds come from external funding sources.

Shenhav also found that the influence of scientific literature on problem choice was inversely related to the "proportion of funding" from external sources and the influence on problem choice of "external suggestions" was directly related to the proportion of funding by external sources.

This result apparently strikes at the heart of the rationalist claim that problem choice in science is based on autonomous scientists freely picking from arrays of previously identified problems. Clearly, Shenhav shows demand from non-scientific quarters can play an important role even among presumably autonomous academicians.[18]

Shenhav's results about external influence may be anomalous; however, I do not think they are. Historians have also added evidence of the importance of "resource dependence" (He who pays the piper . . .) to problem choice. Westfall (1977:111) provides an interesting example. He notes that as early as the seventeenth century, the Academie Francaise des Sciences, whose scientists "were the best equipped in Europe, sponsored a project which succeeded in measuring the length of one minute of arc on the earth's surface, determining the size of the earth with great accuracy."

Westfall (1977:111) insists, according to Shenhav (1985:22) ". . . for these [research] benefits a price was paid. Since [the Academie] was financed by the French government it was also commanded by the government . . . and it imposed activities on scientists who might have occupied themselves otherwise if left alone."

Barnes (1974:121), Shenhav indicates, argued that the distinction between science and its environment makes little sense the further one goes back in history and Shenhav explicitly limits his theory of resource dependence to circumstances where there is clear differentiation of science from the general culture (recall Marx's comments about his performing scientific research [*supra*]), and institutionalized collegial control within the scientific community (Shenhav 1985:34). Besides the challenge to rationalism from some historians, work in two well established traditions of sociological research and theory also suggests that the assumption of the rationalist model about researcher autonomy in problem choice is questionable.

Within organization theory, according to Shenhav (1985:29), various theorists and researchers working in the tradition of "Resource Dependency Theory of Organizations" (Thompson [1967]; Jacobs [1974]; Aldrich and Pfeffer [1976]; Pfeffer [1972]; Pfeffer and Salancik [1978]) argue that:

> . . . elements in the surrounding environment have the power to influence the organization's behavior and to shape its activities, due to the organizational dependency on external resources. Organizational participants try on their part to reduce this dependency (or its inter-related uncertainty) by formulating strategic responses.

A similar argument is raised by social exchange theorists (Blau [1964]; Ekeh [1974]). White [1992] can also be counted among this group of theorists in his major statement of the links of network analysis to other theoretical traditions in social science.

White (1992:76-77) provides an example of efforts by resource dependent researchers to control the environment to reduce dependence (drawing on data from Susan Cozzens (1989)). He relates that in a ". . . controversy over a multiple discovery in neural pharmacology. Five or so distinct research groups are struggling for some recognition as initiators, and many other groups and isolates are interested bystanders and contributors." White says that "stories both of honesty and of originality are being negotiated in interaction with one another through various agencies in a complex field." Implied in his account is that the struggle is an attempt by each team to win resources for the future. The outcome of being recognized as a "co-discoverer" of the 'opiate receptor' is (a) the ability to bargain more effectively with funding sources because of the prestige (White refers to it as "adjustment of standing") accompanying recognition of one's priority and (b) access to other funding sources one has not up to now been able to tap for funds. Thus, his case example illustrates that battles over "priority" are strategic efforts of resource dependent research teams to reduce resource dependence and enhance their own power.

One other study, (Velho [1990]) is apposite here. In her research on problem choice by Brazilian University-based agricultural scientists, Velho found (1990:511) ". . . wide agreement amongst the informants to the effect that financing agencies have very little influence on their decisions about research topics."

She added that "this fact not withstanding, research funding agencies do have a kind of veto power, and might be expected to influence the research decisions made by a scientist. This veto power, however, is exerted first and foremost by the members of the scientific community and tends to reflect their own interests."

The Constructivist Model

The second major line of criticism directed at the rationalist perspective comes from the social constructivist school. The social constructivist perspective emphasizes that scientific knowledge, indeed, all knowledge, is a social construct. Though its immediate roots are in the phenomenological school of philosophy of Edmund Husserl[1], social constructivism's roots can be traced back to Marx's comments (*supra*) regarding science, especially his observation that its concepts, in fact everything about it, has a social basis.[19] Zeldenrust (1990:8) describes the principal features of the social constructivist approach: (1) "non-rational" and (2) "nonhierarchical."

The social constructivist approach is "non-rational" to the extent it emphasized that problems are not chosen "from a set of well established alternatives." The approach emphasizes the "construction of problems" as a social activity.

Secondly, it is "non-hierarchical "because of its emphasis on "local conditions as prime movers" in problem selection. For example, the social constructivists point to "available resources, such as substances or instruments" (or combinations of substances and instruments) as "originating new choices.

Knorr refers to combinations of substances and instruments as 'assets' that can be "used by the scientist to access a specific perceived 'demand'. Therefore, it follows that the asset is "the perceived solution" to "someone else's problem."

The social constructivists thus directly challenge the major feature of rationalism: the nature of problem choice itself and the idea of an "array of previously identified problems as a source of problems scientists investigate."

However, another feature of the constructivist perspective is noteworthy in connection with the present study. The constructivist view suggests a "decoupling" or loose coupling of demand (for a solution) and problem. A problem can be found in the constructivist perspective, in response to someone else's "demand" (as perceived by the scientist). The demand takes the form of a problem looking for a solution.

Here we have, as Zeldenrust points out, (1990:9) evidence that in contrast to the rationalist perspective the constructivists provide a "separate role" for "demand (and resources)" apart "from research problems". This is indicated, Zeldenrust argues, because of the stress in the constructivist perspective on the "significance of" the opportunities to which research selections are addressed." Both Knorr (1979; 1982) and Latour (1980), Zeldenrust asserts, speak of opportunism in scientific problem selection implying "preferences which depend on ... perceived opportunities" and "vary with [those] perceived opportunities."

While the constructivist view is inconsistent with the rationalist view on some points it aligns nicely with the resources dependent perspective. For instance, consider the issue of intentions.

Intentions are largely absent in the rationalist perspective because the 'invisible hand' of 'theoretical needs' is the driving force. (However, in the Lederberg-Zuckerman-Fisher version theoretical need is not the only criterion. Among problems those highest in overall utility on assessed theoretical, organizational, and political-legal feasibility are selected. This does not, however, require intention as a concept.)

In the constructivist view, intention plays a *small* role. It is subject to *change* in the face of changing opportunities - (a big paycheck for doing something the researcher never otherwise would have thought worthwhile, for example). However, intention is needed because the researcher is involved in negotiation of some sort to make the problem technically *feasible* or to permit the researcher to do something intrinsically interesting from the researcher's standpoint as part of a project the client wants to have done.

Furthermore, the constructivist view makes no assumption that problem selection is in the hands of the scientist. If control of problems is in the hands of the client rather than the scientist, the scientist, in the constructivist view, responds to the changed "opportunities" with changed intentions.

It is important to add here that this implies that social constructivists see problems as *locally* originated, *i.e.*, a consequence of specific laboratory conditions and actions. This directly challenges the rationalists' assumption that theory commitments and methodological commitments are central to the process of problem choice. Some studies support the constructivists contention.

Velho (1990) states that her Brazilian study population developed problems locally. Their graduate training abroad taught her study population members to think about issues in certain ways they might not otherwise have approached them, *i.e.*, more theoretically, more in terms of building and testing models; however, the problems her agricultural scientist respondents developed were still locally originated and not drawn from the foreign literature. The laboratory case studies also suggest that problems are locally originated in the specific cases these sociologists (Latour and Woolgar [1979]; Knorr-Cetina [1982]) studied.

The constructivist tradition is strong in Europe, where state intervention in science has a long history (for example, the French Academy of Science during the monarchy, *i.e.,* prior to 1789, had power over projects of a scientific nature as indicated by Westfall's observations, reported above). Shenhav (1985:23) writes that when the French government was ignoring science in the mid 1860s "most French scientists" hoped for "government intervention in science." Shenhav continues, in 1868, in *Le Budget de la Science* Louis Pasteur wrote of the 'suffering' and 'penury' of the deprived community for which he considered himself a spokesman. He described the scientists of France with no financial support for their experiments as soldiers without arms (quoted from R. Fox, [1976:9-10].

Shenhav (1985:23) also indicates that the British government in the same period "... was patronizing some forms of scientific activity" which they justified in terms of commercial military and national benefits."

Shenhav (1985:24) is cautious in his interpretation of these historical data. He concedes that "the exact influences of those patrons on the development of science are not fully understood ...."

Yet, it seems obvious to me, returning to my main point, that with a long history of state intervention in science in France and the United Kingdom, social scientists from those countries would not assume, as the American exponents of rationalism do, that scientists control their problems. Moreover, the European scholars are not troubled by this fact.

Though I believe it is correct in (a) its challenge to the rationalist perspective's assumption about the locus of control of problem choice, (b) its permitting a role for demand and for intention in problem choice, and (c) its recognition of the unstable and opportunistic nature of choice; the constructivist tradition has numerous shortcomings of its own. For example, R. Giere (1988:131) points out that

> ... an excessive concern with current scientific conclusions, especially those in highly controversial areas, makes the constructivist account seem better than it is. It misses what scientists take for granted, namely, the role of previous findings in current research, particularly findings that have made possible the instrumentation for current research. Stephen Cole (1992), though more willing to see merit in the constructivists' insights into how "frontier" science is validated at the outset, also echoes Giele's position about the "core" of a discipline's knowledge (Cole [1992:39-47]). (Cole's entire book is an attempt to assess the contributions of the constructivist scholars within the sociology of

> science; he finds much to like in their specific case studies even though he does not share their view that they have demonstrated that all knowledge is in fact a social construct, as H. Collins alleges (quoted in Gieryn [1982]; see Cole [1992:36]).

It should be clear from all of the foregoing that the three models, the "garbage can", the rationalist hypothesis and the social constructivist perspective cannot all be correct. The rationalist model and the social constructivist model disagree over whether problem choices are locally originated in specific laboratories or reflect commitments to a shared knowledge base and research agenda. The "garbage can" model and the social constructivist model disagree as to whether or not the process is "chaotic." The social constructivists insist that it is because choices arise in the interaction of unique laboratory conditions, experimenter actions, *etc.*. The "garbage can" model acknowledges that the choice is unique but that there are clearly predictable elements in the choices to which the researchers are committed (as suggested by Zuckerman from her review of the "internalist" literature).

There is evidence in the literature to support all three views. For example, Gieryn's study (1978) of astronomy dissertations and the social science literature referred to by Zuckerman (1978), based on research in physics and astronomy, suggests that problems are drawn from the identified stock of problems which can be related to current paradigms; literature cited by Zeldenrust (1990), Knorr-Cetina (1995), *etc.*, drawn from studies in biotechnology, however, suggests that problem choices are locally originated and have at least some unpredictable elements.

My own research findings suggest that the constructivist claim that problem choices are local is not generally correct. Local origination is hardly found in some disciplines, especially the behavioral, economic and social sciences (and probably the physical sciences though my own data are too thin to assess this). However, I found local origination of problems common in medicine and related biological sciences. (See Table 5.2, Chapter 5). Such a finding calls into question *both* the rationalist hypothesis *and* the social constructionist model.

Another finding of my study that calls into question the rationalist model related whether the problem chosen was derived from theory against whether comparisons of problems were done before choosing a problem. Because such comparisons were more common when the problem chosen was derived from theory, it cannot be presumed that problems are shared. It is more likely that they are shared if the problem was derived from theory and less likely to be shared if the problem was not derived from theory.

A further finding from my research is apposite here. After I had noticed that the answers to open-ended questions seemed to be different for men and women respondents, I produced several cross-tabulations of gender by a variety of problem choice related variables. I found a significant relationship between gender and problem choice criteria (*see* Table 5.19, Chapter Five). The social constructivist perspective would not predict gender based differences in problem choice criteria and the rationalists could only explain it by sharply revising their assumptions. Furthermore, other findings on the relationship of problem choices to organization traits similarly did not fit social constructivist perspectives and would require changes in the rationalist model. Together, my results suggests that a different model, the "garbage can" perspective of Zeldenrust, which combines some features of the other two, might be appropriate.

Zeldenrust's own study illustrated the possibility of applying a "garbage can" approach to problem choice but did not constitute a test of his model. Leaving aside a citation study he also performed, Zeldenrust had only archival data, supplemented by interviews of informants, regarding seven research units in Dutch universities. (The units were quite stable over time; they had existed in some cases, albeit with turnover of key members, for a period of some 15 to 20 years by the time Zeldenrust studied them.)

However, an interesting, but neglected, study [not cited by Zeldenrust] by Busch, Lacy and Sachs (1983) provides powerful empirical support for Zeldenrust's general model in my opinion. Their study was inductive in nature and exploratory, seeking to find factors affecting problem choice in agricultural science.

After developing a list of 21 criteria of problem choice through interviews with agricultural scientists, Busch, Lacy, and Sachs distributed a nationwide mail survey to a "random sample of agricultural scientists who were principal investigators in publicly funded institutions (1983:192). Altogether, "completed questionnaires were returned by 1,431 scientists for a response rate of 76.3 percent."

The results reported in the article, according to the authors, are based on responses of their study population to one question: "During the last 5 years, how important were the following considerations in your choice of research problems?" (*loc. cit.*) Busch, Lacy, and Sachs state that "Scientists report that they choose their research largely on the basis of their personal desires" (p. 193)."

This result conflicts with other research (*e.g.*, Nederhof and Rip, n.d.) which makes research problem choice an action that is socially influenced. I subscribe to the latter view and believe that the fact that Busch, Lacy, and Sachs asked for recollections going back up to five years may account for their finding, which is at variance with my and other researchers' views.

The most interesting findings, however, result from their factor analysis of the 21 criteria which resulted in a reduced number (five) of factors. Their table 2 is reprinted here (below see p. 43) to facilitate the reader's following the analysis.

The first factor, which they label "Client Orientation" is clearly a nonscientific audience demand factor. For example, "extension feedback," "clientele demands" and "client needs" load heavily on this factor.

The second factor "peer approval" is clearly a scientific audience factor. Such variables as "colleagues' approval" and "evaluation by scientists" load heavily on the second factor.

Busch, Lacy, and Sachs found that demand from client and scientific audience were "at best unrelated" and perhaps even "in opposition" to one another (1983:196).

The third factor, "scientific ideals," is a problem factor, *i.e.,* the characteristics of the problem itself motivated the interest in working on the problem. This is inferred from the high loadings of criteria such as "scientific curiosity" and "enjoyment."

Table 2.1: Promax Rotated Factor Pattern of 21 Criteria for Research Problem Choice

	Orientation	Client Approval	Peer Ideals	Scientific Advance-ment	Career Utility
Extension feedback	.87	.08	.08	-.10	-.17
Clientele demands	.76	-.02	-.08	.01	-.02
Farm/industry journal publication	.72	.04	.07	.14	-.07
Experiment station Publication	.71	.02	.03	.17	-.08
Client needs	.69	-.11	.01	-.12	.22

Colleagues approval	-.01	.89	-.04	-.12	.07
Evaluation by Scientists	-.09	.83	.08	.03	.01
Credibility of Others	.09	.72	.15	-.02	-.04
Hot topic	.09	.48	-.21	.17	-.01
Scientific curiosity	.00	-.06	.80	.01	.01
Enjoyment	.15	.10	.79	-.07	.01
Scientific theory	-.34	.04	.48	.19	.06
Funding	-.05	-.09	-.10	.78	.02
Research facilities	.19	-.18	.32	.63	.02
Professional journal Publication	-.11	.15	.17	.61	.02
Time	.09	.16	-.19	.60	-.06
Research organization	.22	.17	-.17	.31	.17
New methods or Products	-.13	-.06	-.03	.02	.84
Empirical results	-.17	.09	.04	.11	.59
Importance to Society	.20	.16	.15	-.20	.54
Marketability of Product	.37	-.10	-.09	.07	.47
Eigenvalues	4.5	3.0	1.6	1.2	1.1
Cumulative % of explained variance	.21	.36	.43	.49	.54

A fourth factor, that Busch, *et al.*, labeled "career advancement" arises because the highest loadings on it are "funding," "research facilities," "professional journal publication," and "time." Busch, Lacy, and Sachs (1983:196) claim that all of these criteria for problem choice relate to practical issues involving the possibility and probability of completing publishable research.

Their interpretation of the fourth factor may be true. But a more straight-forward reading, in my opinion, is that this is a "resources" factor because all of the criteria relate to the need for resources to do research. Thus, even "professional journal publication" is a resource-related variable because publication in an appropriate journal enhances one's recognition in the profession and such recognition in turn facilitates acquiring research funds and real resources.

The fifth factor "utility" suggests a dimension that may reflect an organizational cultural variable. Thus, "new methods or products," the highest loading variable suggests that agricultural scientists giving this answer are involved in working for employers who place a premium on *practical* results, *i.e.,* employers who want the researcher to do *applied* research with a good possibility of leading to commercially useful products. All of the other variables have this "results oriented" aspect suggestive of an organization culture variable with a decidedly programmatic orientation. Even "importance to society" which casts the issue in lofty moral terms suggests this interest in research that addresses important *needs* of people and leads to results likely to meet those needs.

The results, then, point to: (a) two kinds of demand, as Zeldenrust found; (b) problems, (c) resources and (d) a factor suggestive of an organizational culture variable specifying that problem choices should be in the applied area. These results are consistent with Zeldenrust, though, I maintain, because Busch *et al.* mislabeled factor IV, I think they failed to notice the relevance of March's "garbage can" model to their empirical findings.

Had Busch, *et al.*, made the intellectual connection between the "garbage can" model and their own work, they would have anticipated Zeldenrust's work, and part of my own, by many years.

The empirical support offered for Zeldenrust's "garbage can" model in Busch, Lacy and Sach's study does not mean that the model, as presented in Zeldenrust's work, is completely developed. Zeldenrust, himself, indicates areas of research with his model that deserve scrutiny. He counts among these "the study of competitive behavior among research groups" (a subject to which I shall return in Chapters Two and Four); about which he provides cursory information. Other subjects that he has not defined as important, I believe, also should be mentioned: the importance of organization culture and of gender of the principal investigator(s) of research teams. These are variables that change the salience of particular flows, and thus the probability of linkages of particular kinds; therefore they need to be explored.

Summary

This chapter's purpose was to present contrasting models of problem choice that might be used as the basis for explaining why men and women scientists choose different problems and more generally why scientists choose particular problems for research and ignore others.

After pointing out that prior research by Fox (1992) and others (Feldt [1985; 1986]; Knorr-Cetina [1995]; Zuckerman [1992];) foreshadowed the finding of gender based differences in problem choice, I presented my perspective on problem choice, based on the "garbage can" model of S. Zeldenrust (1990 Ph.D. diss.). I contend that this is the most promising approach to understanding gender based difference, in problem choice and also in developing a comprehensive model of problem choice in science.

The main feature of this model is its view that such choices are made in conditions of ambiguity. Consequently, the choices are seen as linkages that occur when demands, problems, resources and (possibly) constraints flow close together. Items from any one of these flows may precede the others in a particular case, *e.g.,* a demand may be the "driving force" in certain organizational cultures and precede the finding of a problem to which it becomes linked. The reverse can also be true -- the problem can precede the demand. The linkage thus is based on what items are present among the four flows in a situation where items can come, go and wait from any of the flows.

After discussing this model, adapted by S. Zeldenrust, based on work by James March and his collaborators, I discussed the principal features of three competing perspectives of problem choice. First, I discussed the main assumption and thesis of Zuckerman's rationalist perspective, showing how it is the culmination of work by others pointing to implications of paradigms as affecting the problem choices of scientists.

I then presented the contrasting views of the resource dependence theorist Y. Shenhav and the social constructivists, among whom I count Knorr-Cetina (1995), Woolgar and Latour (1979), Latour (1980), and H. Collins (1982).

The view of the social constructivists and the rationalists are mutually exclusive in two related areas: (a) whether problem choices are locally originated or derive from theory commitments and methodological constraints, and (b) whether the choices have predictable elements, as suggested by the rationalists, or are "chaotic" as suggested by H. Collins and others working in the social constructivist tradition.

As I indicated above, there is literature, including my own findings, to support both positions. Thus, each seems to have some value but neither is a complete substitute for the other. (The resource dependence perspective also discussed, does not imply a dynamic for problem choice. At most, it suggests limitations on the possible choices because of the veto power of the resource provider.)

The "garbage can" model rejects some elements of each of the other two and embraces others. For example, the "garbage can" model rejects the extreme social constructivist position that everything is a social construct and that no knowledge is more correct than any other knowledge. (More moderate statements of the social constructivist position such as Knorr-Cetina's recent writings (1995), though permitting a distinction between "core" and "frontier" are silent on how the transition occurs because the social constructivists emphasize (a) only the social content of knowledge and (b) the "chaos" of the process of selecting research problems. This is because the social constructivist view provides no place for *cognitive* content in its formulation.)

The "garbage can" model rejects the rationalists' *assumption* of a previously identified set of from which a problem (or problems) is chosen for research. For the "garbage can" model, it is an empirical question where the problem was found. However, the source of the problem is not especially interesting.

Even though the "garbage can" model does acknowledge that problems can be derived from the previously identified stock of problems, unlike the social constructivist model which insists that all problems are locally originated ("negotiated" in the unique circumstances of particular laboratories, particular social actors, *etc.*), it is evident why Zeldenrust's perspective is neo-constructivist; there is an obvious intellectual debt to the constructivists in his position that research problem choices are opportunistic; *i.e.*, the choices occur when a linkage is made among four loosely coupled "flows." (And a linkage is clearly "the unintended consequence of interacting actions.")

But aside from Zeldenrust's insistence on using stable research groups as units of analysis, what is specifically *organizational* about his model? The answer lies in its emphasis on attempts by the unit to control its environment. Acknowledging his debt to the resource dependence theory of Pfeffer and Salancik (1978), Zeldenrust maintains that organizations need to "reproduce control over [their] 'environment'" and they "usually desire to increase it." Also important, but less emphasized in writings about this issue is that the control over the environment "is not an end in itself" but stems from "the need for control over the organizations' own processes (*e.g.*, through growth)."

Zeldenrust then adds, for the problem choice process of interest here that "the research group as a 'local hearistic organization' managing 'the interface of choice arena and environment' control over the problem choice process refers to the organization's "grip" over the "the parameters" and the "configuration of the process" which in March's terms translates into control of "the contents of each of the flows" and of the "inter-relationships" of these flows.

In the next chapter, I shall set forth a strategy by which the "garbage can" model can be assessed empirically in five areas: autonomy, derivation of problems, competition, organization culture and gender differences. I also will indicate why there are useful areas of inquiry. First, however, I want to summarize the main characteristics of the three current models discussed in Chapter One and the "garbage can" model, as presented by S. Zeldenrust. This I do in Table 2.2.

Table 2.2 Comparison of Rationalist Models and Other Perspectives

Dimension	*Rationalist Model*	*All Other Models*
Where is locus of control in choosing problems?	Assumed to rest with scientists themselves, i.e., *scientists* make decisions	May rest with clients (resource dependency theorists, social constructivists). *Organization* makes decisions (neo-constructivists)
How are problems selected?	From stock of problems developed by the leading researchers in field through comparison of problems.	Developed locally (social constructivists). However, as indicated below, this study maintains that problems can come from anywhere.
Do social factors enter into problem choice?	*Possibly*. In the Lederberg-Zuckerman-Fisher formulation this occurs during the problem comparison process. In the early Zuckerman version (1978) little likelihood for this external	Yes, *definitely*, (constructivists, resource dependency theorists). Different theorists point to different mechanisms by which social factors can influence problem choice. Example,

	influence to have an impact appeared to exist.	Knorr-Cetina points to "transepistemic arenas" (1982).
Major weaknesses of model:	1. questionable assumption of researcher autonomy in problem choice. 2. cannot adequately account for local origination of problem which is common in some disciplines. 3. counterfactual implication of widespread interpersonal competition in science. 4. questionable assumption that problems are "shared"	1. too much emphasis on local origination of problem in social constructivism; in most disciplines there is far more derivation of problems from theory *etc.* than critics of rationalist perspective believe. 2. failure to see underlying regularities in problem choice (social constructivists) 3. lack of satisfactory means of explaining control strategy formation (Zeldenrust model).

ENDNOTES FOR CHAPTER TWO

[1]Quoted in Rose, Richard (1976) "Disciplined Research and Undisciplined Problems," *International Social Science Journal*, 28:1, pp. 99-121 reprinted in Chubin, Daryl E, Alan L. Porter, Frederic A. Rossini, and Terry Connolly (eds) (1986).

[2]Strictly speaking, the choice is made when a demand and a problem are linked in a context where resources and constraints are also linked together with the problem and the demand. However, many times a preliminary linkage is made and then, in Zeldenrust's terms, "unblackboxing" is needed in order to make the complete linkage among demands, problems, resources and constraints.

[3]Zeldenrust's model is one of a class of formal organization models generated from the "garbage can" perspective of James March and colleagues, originally developed by them to understand universities (which March calls, oxymoronically, "organized anarchies").

[4]The actors in Zeldenrust's model, called by him a "local heuristic organization," are biotechnology research units of analysis. The purpose is to remove "researchers as individual actors, their motivations, socialization, or job mobility," because he believes a focus on the individual researcher obscures higher level processes that account for *regularities* in research unit behavior. As he puts it (1990:5), "individuals are mere appendages of the choice process. Their shells are resources for the local heuristic organization." Zeldenrust's model, therefore, is a model of organizational decision-making whereas both the social constructivist and rationalist models are models of individual scientist decision-making.

[5]White's (1992:52-54) concept of "arena" seems to be essentially the same. White says of arena that they are "a sift-and-exclude" species which certainly would be true in the case of a research team as "local heuristic organization."

[6]A famous example of a "garbage can" model is Kingdon's (1995; originally 1984) study of policymaking in Washington D.C. at the federal government level. The "garbage can" model was first described in Cohen, M.D., March, J. G. and Olsen, J. P. (1972) "A "garbage can" Model of Organizational Choice." *Administrative Science Quarterly* 17:1-25, Also see March, J. G. and J. P. Olsen (1976) *Ambiguity and Choice in Organizations* Bergen etc: Universitetsforlaget.

[7]Zeldenrust defines a research problem as "the substantive objective in pursuit of which . . . experiments [or 'thought experiments' or quasi-experiments] are [undertaken, and it is the direction they take. A research problem involves a design: what is the question, how can it be operationalized, which data can and should be collected . . . [etc.]." And, he adds, a problem is ". . . an objective orienting action [that] *creates* [locally] an area of ignorance, rather than being derived from it."

This is not, *per se*, inconsistent with a rationalist perspective. A problem created by a general theory can also become an element inserted into the problem search-process of the research team. It then (based on the elements in the situation, including the problem) is linked to a demand to create the choice. The point is that, in the Zeldenrust model, it does not matter how the problem is found.

[8]Zeldenrust considers the research choice to be equivalent to a *linkage* between "a demand and a problem." He states that the process of linkage is "unconstructed" because it "consists of the unintended consequences of interacting

actions by various largely 'invisible' actors (both inside the group and in its environment) who, knowingly or not, insert, remove or change elements in the choice arena. This linkage, he adds in a revealing footnote (1990:4n), is "similar in principal (but much more unpredictable in practice) to a market mechanism, where prices are the unintended consequence of actions of buyers and sellers."

For the rationalists, demand is presumed, immanent in the problems, at least in the significant problems, and not an object of analysis. The rationalist model's conflating of problems and demands causes a paradox to arise. As Zuckerman (1978) observes, there are premature mature, and (presumably) post mature discoveries in science because she presumed that problems generate demand for their solution. Post mature discoveries, thus, present an awkward situation for the rationalists. Why, if problems are known to be important and the intellectual tools exist for their solution, should problems not be addressed and solved promptly? Delayed attention to problems suggests some distortion or deviant social process interfering with the natural progress of science.

The "garbage can" model resolves the paradox simply and elegantly. Since *demand* is a distinct flow in the "garbage can" model, and since demand is culturally patterned, one can understand that foci of research are not coupled to the solvability of the problems and thus problems are not prematurely or post maturely addressed.

[9]In my opinion, Shenhav is only partly correct. The idea of science as autonomous is, indeed, an aspect of Parsons' work. However, Parsons does *not* require absolute autonomy for science. For instance, he says (1951:343) that the scientist role "is of the special type . . . called a professional role." The professional role is a position filled by persons who have specialized knowledge and who must be given some freedom from outside control; because as Parsons observes [1957:335], "the knowledge [the scientist] possesses is only with difficulty if at all accessible to the untrained layman, the ultimate judgment of it must lie with his professionally qualified peers."

[10]The rationalists see autonomy as a reasonable presumption. Percy Cohen (1977:64), a British social theorist, puts the rationalist view succinctly when he says ". . . a definable part of the social system [*i.e.*, science] can have a high degree of autonomy with respect to others but at the same time influence and be influenced by them."

The rationalist scholars are well aware that science is not linked to other institutions of society in this manner everywhere; for example, they would acknowledge that the autonomy of science in "Communist" block countries prior to the breakup of the Soviet Union in 1991 may not have been nearly so considerable as in the United States and other western "democratic" nations. However, the rationalists, pointing to the damage done to Soviet genetics research by Trofim Lysenko as an example, would undoubtedly argue that autonomy for scientists is in the best interests of both science and society.

[11]Debaekere and Rappa (1994) did a mail survey study of "problem choice behavior" among "neural network" scientists (a branch of computer science) worldwide. Of the 1875 scientists to whom questionnaires were sent, 452 (24%) returned useable responses. Debaekere and Rappa (1994:438) found that academic institutions have "special significance"

> . . . in fostering the development of new fields of science and technology. To the extent that scientists are provided the freedom to interact with their colleagues and pursue their interests with singleminded determination, early entrants to a field are perhaps more likely to be found in an academic context.

They add that ". . . early entrants in industry (and to a lesser degree, their government colleagues) are not unlike other scientists who enter after them . . . ."

This latter finding Debaekere and Rappa (1994:438) saw as "an indication of how the special constraints placed on non-academic scientists may affect their choice of problems."

Considering the passion with which academicians defend "academic freedom," it seems a bit surprising that just one study (possibly biased by sampling attrition) could be found attesting to the value of giving scientists free choice in problem selection.

[12]Zuckerman assumes that research problems of assessed importance create their own demand for solutions. The reasoning is that scientists, seeking priority for important discoveries, will flock to such problems, even competing head-to-head to solve them first. Recently, however, (Stephan [1996]) has questioned this view from an economist's standpoint (1996:1225).

> The conventional wisdom holds that because of problems related to appropriability public goods are under produced if left to the private sector. Although priority goes a long way toward solving the appropriability problem in science, this ingenious form of compensation does not ensure that efficient outcomes will be forthcoming. In addition to problems caused by uncertainty and indivisabilities as well as other efficiency concerns.

> . . there is the problem that scientific research requires access to substantial resources.

The most interesting of Stephan's list of hindrances to researching problems of significance, from the standpoint of this study, is "uncertainty." This may include uncertainty about the importance of a problem, *i.e.*, the problem's importance may not be understood by all before the study is undertaken and, even afterward, for a great length of time. Thus, it is not at all evident that scientists will be drawn by "priority" for their work to study the important question.

[13]The discovery of the phenomenon of sonoluminescence is a case in point. Though scientists have known for decades that sound waves can excite little bursts of light when the sound bombards a liquid such as water, the phenomenon was regarded as a mere curiosity until recently. New appreciation of potential commercial/military application has drawn more physicists to study this phenomenon. (See *New York Times*, September 16, 1997, C1.).

[14]Shenhav (1985:1) notes different scholars have opposing views of which are external and which are internal factors, and differentially weight their relative roles in the explanation. If scientific investigation is perceived as an entirely rational-logical process as the philosophers, and some of the historians, define it, then any social process is external to the scientific endeavor. Shenhav points out that "most sociologists of science" accept the view that social processes internal to science are differentiable from processes external to science and that when the former are *exclusively* the factors used by a sociologist to explain phenomena in science, the argument is an internalist argument.

I endorse a view that "external" factors are those outside the boundaries of science as an institution. For example, businessmen's needs are "external." However, the businessmen's needs may be "translated" into terms understandable to scientists by scientifically literate nonresearchers in what Knorr-Cetina (1982) calls "transepistemic arenas" and the influence of such on science is more difficult to determine as clearly external or not.

[15]See Kuhn, "Postscript - 1969," in *The Structure of Scientific Revolutions, 2nd ed., enlarged*, for Kuhn's attempts to answer critics and indicate refinements in his own point of view.

[16]My position is based on the interpretation I give to Leon Festinger's cognitive dissonance theory. (Festinger 1957). The main argument of Festinger's theory is that once a decision is made, arguments that seem to call that decision into question are suppressed because of their capacity to cause psychological dissonance, an unpleasant emotional state. Dissonance among researchers may be created, I hold, by arguments against a particular theory or experimental design, etc., once a decision to use that (those) concept(s), paradigm design(s) etc. has been made. And similarly, questioning the "scientific importance" of the projected

research after it has been decided upon would be dissonance inducing. Therefore, it seems plausible to argue that the decision to do the research could heighten the "assessed scientific importance."

Besides the methodological problems that must be resolved whenever there is a need to model two-way causation--and determining the order of precedence between the "decision to study an issue" and the "assessed scientific importance" of that problem as a research subject seems to be an example of two-way causation--I have other qualms with a model of average scientist behavior that uses "assessed scientific importance" as a predictor variable. Operationally, for example, how is "assessed scientific importance" to be defined? Can anyone decide this or only certain researchers? If the latter, what criteria should be used to determine suitable raters? (It is important to note here that Lederberg and Zuckerman do not address this problem. However, it is not necessary for them to consider this problem in the special case of elite scientists they were concerned with because elite scientists clearly are suitable raters. It is quite different in the case of average scientists for whom I am interested in developing a model of problem choice.)

I do not know that the problems of the concept of "assessed scientific importance" are insoluble but the term does present practical difficulties I prefer to avoid.

[17]Graham's views are not shared by all writers. Raissa Berg (1990) a Russian geneticist, in a personal memoir characterized Lysenko as "an all-powerful charlatan" in contrast to Graham's more measured view of his power and competence.

I see organization culture as playing an important role in problem choice. In the Lederberg-Zuckerman-Fisher version of the rationalist perspective, the organization culture specifies the values against which a potential problem is weighed to determine its "organizational feasibility." For example, feasibility is threatened in Catholic hospitals when research is thought to require tests of drugs interfering with the development of a fertilized egg into a child since life in the Catholic faith begins at conception.

In the organization studies conception I now endorse organization culture is important in creating the control strategies that guide problem selection. I follow a definition of "organization culture" given by Sonja A. Sackmann. She says (1991:34) that organization culture refers to "sets of commonly held cognitions that are held with some emotional investment and integrated into a logical system or cognitive map that contains cognitions about descriptions, operations, prescriptions, and causes. They are habitually used and influence perception, thinking, feeling, and acting.

[18]In Chapter 3 I discuss the methodological inadequacy of Shenhav's research as a falsification of the rationalist perspective. However, in this chapter I am concerned only to demonstrate the existence of rival hypotheses to the rationalist perspective and to adduce the evidence on their behalf.

[19]"It is not only the material of my activity - like the language itself which the thinker uses - which is given to me as a social product. My own existence is a social activity." Marx (1964).

CHAPTER THREE: A STRATEGY FOR TESTING THE "GARBAGE CAN" MODEL OF PROBLEM CHOICE AGAINST RIVAL PERSPECTIVES

Prefatory Remarks

In the first chapter, after reviewing four models of problem choice -- (1) the rationalist; (2) the resource dependence; (3) the social constructivist; and (4) the "garbage can" model -- and indicating their major assumptions, principal hypotheses, and findings from the literature bearing on each, I concluded that the "garbage can" model seemed more suitable as a basis for an empirical theory of problem choice than any of the others. This is a desirable outcome for my purposes because, while I do not need to explain gender differences in problem choice (my secondary objective in this study) by reference to any existing models of problem choice, I would like to tie my explanation of such gender based differences to a larger model of problem choice.[1] Furthermore, I prefer to have the "garbage can" model as a basis for an empirical theory of problem choice because not only does it seem superior to the others in general, but specifically for the narrow purpose of explaining gender differences it is more useful than the others. The reasons for the superiority of the "garbage can" model for the purpose of explaining gender differences lie in the assumptions and hypotheses of the respective models.

The "garbage can" model is more parsimonious than the rationalist model because unlike the latter, the "garbage can" model does not presume researcher autonomy or prior knowledge of the importance of problems. (The other two models discussed, the social constructivist and the resource dependence, likewise make no presumptions about autonomy or prior knowledge of problem. However, they have numerous other drawbacks).

There is one further advantage to the Zeldenrust "garbage can" model from a theoretical standpoint. Its emphasis on the organization's striving for control of its environment makes this model consistent with other traditions in organization theory, especially inter-organizational models. And the model also seems consistent with individual based models such as social exchange theory and newer models such as a major statement of network theory (White [1992]) that places great importance on the "control struggles" that occur within ties of "identities."[2]

The study posed the question at the outset of why, if problem choice is rational behavior, are there gender based differences in problem choice? From the standpoint of seeking an answer to that elementary question, only the rationalist model and the "garbage can" model are promising. The social constructivist model, asserting that problem choice is local, and "chaotic," is not able to account for any underlying regularities in the choice process, including gender based differences, very satisfactorily. The resource dependence perspective likewise has nothing to say about this issue.

The difference between the rationalist approach and the "garbage can" approach to gender based differences essentially boils down to the following:

The rationalists assume that gender based differences will disappear if constraints imposed by society differentially on women and men scientists were removed. That is, if men and women scientists were equally free to select problems (*i.e.,* subject only to appropriate peer pressure of their colleagues), they would select similar problems.

The "garbage can" model begins with the definition of a problem choice as a *linkage* between two streams: "demands" and "problems" (and sometimes a third stream entitled "constraints", referring to certain special criteria a problem's solution must meet). It then argues that social forces, *e.g.*, organizations and, perhaps, gender differences *independently* (or jointly or both) are important in moving the "demand" stream and the "problem" stream close together and enabling a linkage to occur. Gender differences in preferences are not presumed to disappear in this model if differential treatment of men and women ends. However, the "garbage can" model neither requires nor excludes gender choice differences.

The critical difference between the rationalist and "garbage can" approaches, in my opinion, is in their view of the role of the organization milieu. The rationalists presume that the organization milieu will be benign and uninvolved in problem choice. Problem choices are made by autonomous scientists, who, knowing the importance of problems beforehand, are drawn to solving the most important of them, creating interpersonal rivalries in the process as each scientist seeks priority.

Zeldenrust's "garbage can" model assigns a critical role to organization milieu. Scientists are not presumed to know the importance of problems. Scientists suggest problems meeting specific organization demands or alternatively the scientists propose problems and search for a demand. A choice is made when a demand and a problem become linked. The decision, however, is an *organization* decision.

Given their view that individual scientists make research problem choices, the rationalist model might explain how gender based differences in problem choice could be expected to arise if there were gender patterned differences in researcher *autonomy*. If, for example, women researchers were less able to choose the important problems than men scientists, then there might be gender patterned differences in problem choice (and perhaps in productivity).

Another way rationalists might account for gender patterned differences in problem choice would be to argue that women, for various reasons, might generally choose to avoid the head -to- head competition expected when researchers tackle the truly important problems and seek quieter if less heady areas for research. This kind of explanation would be consistent with the Cole-Singer (1991) thesis of limited differences.

Rationalist explanations of gender differences in problem choice are most appealing when differential barriers to scientist autonomy are found. When such barriers are not so evident, when the assumption of autonomy of scientists *generally* seems questionable, the rationalist approach is not as appealing as the “garbage can” model. That is, while explaining gender based differences in the absence of barriers may not be an insuperable problem for the rationalists, I find their approach less attractive than a model such as the “garbage can” perspective that easily accommodates gender differences without the need for *ad hoc* explanation. The “garbage can” model’s ability to deal with gender differences is the reason it is the approximate answer to the right question, as urged by John Tukey, whom I quote on the title page of Chapter One.

“Approximate” in this connection is used deliberately. For, while I prefer the “garbage can” model to either the rationalist or the social constructivist models, I recognize that, without data on various issues, *e.g.*, what role competition and conflict play in problem choice, or what impact the organizational milieu has on problem choice, it remains just another interesting model, a mere basis for conjecture, however credible.

Therefore, I must turn to the task of how I will show that (a) there is a plausible reason for gender differences and (b) this can be related to a new paradigm of decision making in organizations referred to as the "garbage can" model.

This chapter will focus on two issues. First, I shall set forth a general strategy for evaluating the different models, especially in regard to their ability to explain why gender differences in problem choice occur. As part of this strategy, I will show why, if data of the right sort are marshaled, the reasons for these gender differences can be seen as most consistent with the "garbage can" perspective and less consistent with any other model of problem choice.

Second, I will state specific hypotheses that, if tested statistically, should yield clear evidence for or against the suitability of the various models of problem choice as bases for a theory of problem choice.

Section One: An Approach to Testing The "Garbage Can" Model of Research Problem Choice

In this section, I discuss a general strategy of testing the "garbage can" model against the rationalist and social constructivist perspectives.

My main contention is that the "garbage can" model is superior to the other two perspectives because overall, it predicts best among them across five important areas:

(a) autonomy of scientists in problem choice

(b) local origination of problems versus derivation of problems from theory

(c) competition and conflict as sources of problem choice

(d) organization characteristics as predictors of the team's problem choice criteria

(e) gender of research scientist team leader/principal researcher as a predictor of problem choices criteria

1. Autonomy

Autonomy is an especially important issue in the rationalist perspective. Though I cannot say that anyone has stated the hypothesis in these terms, the rationalist hypothesis essentially says that, "Given freedom to choose, scholars will select the most important problems they can from the identified stock based on theoretical and/or methodological commitments."

For rationalists, autonomy has a particular meaning. It is *not* liberty to do as one pleases. Rather it is professional autonomy. Scientists are assumed free to choose problems without interference from nonscientists. Scientific peers are *expected* to influence the problems chosen by scientists.

Rationalist scholars certainly know that nonscientist influence occurs in problem choice. However, they see this as largely absent from academic science and, when it occurs, they regard it as a negative influence. This opinion of outside influence stems from their conviction that nonscientist influence is likely to lend to problems being chosen that are "less important" (or even in some cases "abused" since they are not "amenable to investigation").

Autonomy of scientists is an assumption based on research in the professions, specifically that showing scientists to be academic professionals. Freidson (1970:134) says that autonomy refers to 'the quality or state of being independent, free, and self directing.' He adds that , "In the case of the professions, autonomy apparently refers most of all to control over the content and the terms of work. ... the professional is self directing in his work."

Freidson makes the additional point that organized autonomy (as opposed to autonomy by default) refers to protection by law of the occupation's autonomy and especially freedom from competition from other workers not recognized as members of the profession.

The professional status of scientists derives from their university affiliation. The sciences are taught in colleges and universities by professors who are regarded as high professionals (*see* Parsons [1951:335-345]. Also see W. J. Goode [1973] for a major theoretical discussion of professionalization).[3] It is reasonable to suppose that the tenured university faculty enjoy considerable autonomy, especially in those areas where little in the way of outside funding is required for equipment, *etc.* However, what about the scientists not affiliated with a university? Much scientific research is done in nonacademic settings, *e.g.*, in military and other government research centers, in commercial, and in nonprofit research centers. A perspective that is applicable only to scientist's work in academic environments may be of limited value in a model of problem choice. (Logically, for a complete model of problem choice we need to know how problem choice occurs in both academic and nonacademic research milieu. As a practical matter, we may only need to know if the proportion of researchers who regard themselves as not autonomous is negligible.)

Later in this chapter, this point will be stated more formally as a testable hypothesis.

The rationalist argument for scientist freedom to choose problems depends on an *assumption of perfect information* about *problem significance*. This assumption is implicit in the idea that problems are chosen from an "identified stock." More will be said about this in the discussion of where problems come from (see below).

The "garbage can" model, as I have already indicated, makes no *presumption* about autonomy of scientists. However, autonomy is more narrowly defined in the "garbage can" model. Autonomy in the "garbage can" model refers *only* to the degree to which scientists are free to suggest problems for study. At one extreme are scientists who labor on problems defined for them by their superiors, who may or may not be scientifically trained. All the others are scientists who, by virtue of their formal position and/or the esteem in which they are held, can suggest problems to be investigated. Even more important these latter scientists are almost certain to receive funding from someone to pursue research on the relevant issue.

Gender Differences in Problem Choice Criteria As a Strategic Research Site for Examining the Rationalist Perspective

As I indicated above, rationalist scholars acknowledge that professional autonomy of research scientists is not necessarily a given in every research center. Professional autonomy is generally found among academic researchers – at least at major research centers in North America.

One interesting argument of the rationalists is that problem choices of scientists who are autonomous will be more similar within a discipline. This assertion follows from their belief that all of them will try to solve the major problems of their discipline.

Implicit in this view about what problems scientists choose is the idea that scientists have perfect information about problem significance.

If the rationalist hypothesis is correct, problem choice differences between men and women should reflect greater constraints placed on the research of one gender (presumably women) than on the other (presumably men). This is the only possible explanation unless one wishes to maintain that one gender is somehow less able than the other to assess the important problems of the discipline. (Aside from being politically incorrect, an argument of gender differences in intelligence is not grounded in any evidence).

There is some empirical support for the argument that women's research is more hampered than men's research (see, for example, M.I.T. Report [1999] which shows women faculty have more difficulty than men getting student research assistants. Also see Feldt [1986] which shows junior faculty women getting less funding support than men junior faculty, as well as Cole and Singer (1991) which suggests that women may have less access to influential senior faculty than men as graduate students and thus fewer women students get appointments to the better graduate departments as junior faculty).

However, suppose we were to find that among men and women faculty reporting similar levels of freedom to choose problems, there were sharp differences in research problem choice criteria (and, presumably, in problems likely to be chosen).

Such a finding would surely raise questions about the factual basis of the rationalist hypothesis and thus of its suitability as a foundation of an empirical theory of problem choice. More will be said about this point in Chapter Five.

2. Theory Derived vs. Locally Originated Problems

The social constructivists generally insist that problem choices are "local," a result of the unique interaction of instruments, demands, scientists, *etc*. However, at least one prominent social constructivist, Knorr-Cetina, allows that "multiple discoveries" can occur in science which implies that problem choices are *not* always local. Knorr-Cetina, (1981:6) quoted in Zuckerman (1988:556), says, "Given that scientists working on a problem are related through communication, competition, and cooperation, and often share similar educations, instruments, and internal structures, the latter situation [multiple discoveries] is not really unusual."

Though Knorr-Cetina might deny that she implies this, it seems that she also allows that problems are often derived from theories. First, she acknowledges that instrumentation can be common to many disparate sites. Instruments are embodiments of theories. However, this is not all. She points to "similar educations" also. This means that she concedes that there can be a core of knowledge,--theories,paradigms, included--from which problems can be derived.

Zuckerman (1988:556) notes that Knorr-Cetina's observation is "not very different from the explanation ... which attributes multiple discoveries to a shared knowledge base and a shared research agenda within the acutely competitive social framework of science."

Though never saying it outright, Zuckerman here strongly hints that she believes that problems are usually derived from *theories*, especially the main paradigms of a field, often embodied in equipment common to many research laboratories and institutes, and included in the "shared knowledge" of the scientists.[4]

There are other hints in Zuckerman's work that she shares Thomas Kuhn's (1970; originally 1962) belief that problems are derived from theories of a discipline. For example, other scholars cited approvingly by Zuckerman (1978) have emphasized the importance of theory, especially paradigms in physics, in influencing what problems are selected for investigation (see esp. Sullivan, Barboni, and White [1978Ms]).

Finally, there is the fact that the rationalist model asserts that "with few exceptions" problems will be chosen from the pool of identified problems in the field.

The notion of an "identified stock of problems" does not in itself suggest that the problems are derived from theory. However, taken together with the other hints in her work, (*e.g.*, her interpretation of Knorr-Cetina's point [1981] cited above), it is hard to accept any explanation for her views other than that she and other rationalists expect most problems to be theory derived.

Zeldenrust points out that in the "garbage can" model the key point to keep in mind is that things are coming, going, or waiting. Ideas, such as theories, can be inserted (coming) in the local choice arena or removed (going). And the same is true of prior findings. Theories and previous findings are not themselves controlling in the process of problem selection. What controls is the totality of items present when a linkage is made between a demand and a problem (and sometimes a constraint).

(It is this phenomenon, I believe, which led to the notion of premature, mature and post mature studies. Scholars who depend on retrospective study, *e.g.*, historians of ideas, would notice that a study could have been derived from a theory--of which the scientists doing the study, perhaps, were unaware--to argue that the study could conceivably have been done much earlier. The present study exemplifies this since the March *et al.* theoretical papers were done in 1972 through 1980. However, I was unaware of their applicability until the mid 1990s).

In conclusion, where problems come from is an issue all the perspectives address with varying degrees of clarity. Only the social constructivist position is really unambiguous about the matter however, while the rationalists are a bit vague. However, there are hints in their writings that they differ sharply on the matter from the social constructivists. The "garbage can" model is neutral; in fact, it implies that where the problems are found is not theoretically interesting.

3. Competition as Cause and Result of Problem Choice

Though competition and conflict are huge subjects that I can only touch on here – and even the specialty of competition and conflict in science includes matters beyond the concerns of this study – they need to be addressed here. However, I am only focusing on the relationship of competition and conflict with problem choice, a topic to which Merton (1973) has contributed valuable insights.

First, let me say that I do not deny the possibility that problem choice(s) can cause competition – indeed, this is an issue R.K. Merton (1957) addressed in his famous essay "Priorities in Scientific Discovery." Moreover, I do not deny that problem choices may even lead to conflicts. However, conflicts and their origins do not concern me here.[5]

Merton is not the only scholar to delve into this issue. In fact, quite the contrary is true. Competition in science has attracted the attention of scholars from many research traditions within sociology and even of scholars from other disciplines including history, economics and political science.

The reason competition and conflict as a result (and as a source) of problem choice needs to be addressed here is not merely the plausibility of this connection but its *importance* in the rationalist research tradition in contrast to the "garbage can" model and other constructivist approaches. One major feature of the rationalist hypothesis is its prediction of widespread intense competition to solve the significant problems of the discipline before others do. In contrast, the "garbage can" model-- indeed all the constructivist approaches-- does not predict widespread competition to solve problems before others do.

Determining if there is a strong relationship of competition with problem choice, therefore, is relevant to my aim of deciding with whether the "garbage can" model of problem choice is superior to rival models such as the rationalist hypothesis.

Zuckerman (1978:83) puts the issue succinctly.

> . . . the process of competition as related to the reward system appears to be attracting the most notice as an influence on problem selection. The social system of science provides institutionalized motivation and reward for achieving priority in solving significant problems at the moving frontier of the field. (Merton [1957] 1973 c, chap.14). This reward system acts to enlarge the numbers of scientists who want to work in the "interesting" and consequential problem areas - the "hot fields," so they are often called - intensifying the competition in those areas. Individual scientists then confront a dilemma: Should they work on problems widely defined as interesting and important, and thereby enter into vigorous competition for priority with many others, or should they select other problems involving less competition?

In this passage, Zuckerman sets forth the most provocative reason for competition among scientists: a limited group of problems known to all. And she also makes clear that the competition is of a distinct sort: "competition for priority."

Competition is, depending on how the term is used, a commonplace, found in all spheres of human activity. Examples of it in science are easy to think of; for instance, scientific honors such as the Nobel Prize, the Lasker Prize, *etc.*, are awarded competitively. Grants for research also are competitively awarded based on generally objective criteria and most likely grant conditions influence problem choices of scientists.

However, Zuckerman is not concerned with the mundane competition for resources under conditions of scarcity that economists address. According to Zuckerman, and others who share her view, scientists are racing to solve important questions before others do, and to claim the honor of being first. This is called "priority" in science. This form of competition only has one *guaranteed* payoff: a symbolic reward, the thrill of knowing you won. Sometimes, fame and even fortune also go to the victor in this race. However, that is certainly not guaranteed.

Let me be quite clear about this issue. No one questions that there is competition in science. However, "competition for priority," a special form of structural competition, is a distinctive concept. The issue of (1) whether there is actually something called "competition for priority" and (2) how common is it are what I wish to investigate.

The argument that Zuckerman makes about competition for priority is so intriguing and persuasive that even scholars such as Mulkay (1991) who do not consider themselves rationalists accept that there really is competition for priority. And scholars who are critical of the rationalist position do not deny that the argument regarding competition for priority may be correct at least in the United States. For instance, Zeldenrust (1990:291) himself, while indicating that he found no competition between research institutes in his Netherlands data allows that perhaps in the United States the situation might be different.

The general acceptance of the "competition for priority" thesis seems odd when one recalls that a fundamental point of the "influential" social constructivist position (see Keith Parsons [ed.] "Introduction to Part 1," in *The Science Wars* [2003:26]) is that the research process is "chaotic." Problems are "locally" originated in this process through the interaction of laboratories, scientists, and other forces.

The reason that some can comfortably accept both a social constructivist perspective and competition for priority is that the social constructivist perspective generally is regarded as applicable to how scientific findings become accepted as facts (see R. Klee "The Sociology of Knowledge: Exposition and Critique," in Parsons (ed.) [2003:57-78]). The social constructivists emphasize that this is a *social* process, and, at least some of the main writers in the social constructivist tradition, according to Klee, deny that the facts are anything other than social constraints. Problem choices, however, are not merely social constructs but real even if knowledge of them is a result of some social process.

Before offering an alternative view of competition in science to the rationalist hypothesis, it is important to restate an assumption implicit in the rationalist view. Zuckerman and others sharing her view assume that there is a set of "identified" problems or "shared" problems (Gieryn, 1978) that can be at least roughly classified by their intellectual importance. If this assumption is challenged, *i.e.,* if it is argued that problem significance is not known beforehand, that, in other words, there is a lack of clarity about the importance of problems, then competition for priority may be much less prevalent within the discipline than the rationalists expect.

An alternative view to the provocative thesis of the rationalists about intense competition for priority is that in an environment where (1) resources for research are scarce, and (2) there is no stock of previously identified or shared problems and/or the significance of problems is not clear to all disciplinarians patterned differences in strategies for obtaining resources will have emerged. Women, for example, may be expected to follow different strategies than men. This is analogous to the differences in strategy for obtaining parental attention used by second born (or generally later born) children than by first born (see Sulloway [1996]). In particular, according to this alternative view, women who have traditionally faced barriers to their entry into certain disciplines, e.g., engineering, and obstacles to gaining access to expensive equipment and supplies for scientific research (see MIT Report [1999]) seek to differentiate themselves from men scientists in notable ways, e.g., (a) women may be more likely than men to emphasize topical relevance of the work to important concerns of society; (b) women may be more likely to make certain that their problem choices are endorsed by renowned scholars and (c) women may be more likely to be concerned about meeting deadlines. These and other differences may be intended to highlight that women's scientific work should be taken seriously and that investments made in women's science will be worthwhile, e.g., will be rewarded by good publications on issues people care about.

The alternative view I propose of patterned differences in strategy in an environment of ambiguity about problem significance leads to *very different* predictions than the rationalist thesis:

- Given a high degree of freedom to choose their problems, women will choose problems different from those chosen by men. (This is quite *the opposite* of the rationalist's prediction).

- Women will emphasize different criteria than men will in problem choice, as part of a strategy to differentiate themselves from men advantageously and thereby encourage investment in their scientific work.

4. Organization Influence on Problem Choice

While demonstrating the wide and deep impact of the work organization (or more exactly, the "organization of orientation") on the problem choice criteria of researchers is necessary to showing the value of the "garbage can" model of problem choice, it does not directly bear on the question of gender differences in problem choices. However, since an argument for the importance of organization influence on the process of problem choice is, broadly speaking, a cultural explanation, it opens the door to other cultural explanations including gender effects (either alone or together with organization effects).

The influence of the work organization on the research criteria of the scientist, and more generally on the research team's organization culture (of which criteria of problem choice are an important component) can explain the occurrence of some puzzling phenomena such as multiple independent discoveries and “post mature” discoveries. The latter is particularly problematic for a rationalist model that argues for demand based on the known importance of problems. It is important to note, however, that Zeldenrust is silent on the process within the "choice arena" of the team by which a choice is made. It can be inferred from the theorizing of White (1992b) that, since choice arenas are a "sift-and-exclude species," it follows that the team requires some value or technical standards as a basis for rejecting problems found wanting before a choice is made of a particular problem (by the matching of a demand, a problem, and resources--and sometimes constraints).

Organization culture of the team is also important in the Lederberg-Zuckerman-Fisher version of the rationalist model (but not in the original formulation of the rationalist hypothesis by Zuckerman) as a factor in problem choice. While the rationalist perspective, even in the Lederberg-Zuckerman-Fisher version, is profoundly different from the “garbage can” view, in regard to the effects of organization influence on problem choice they may not yield dramatically different predictions. (The differences are much clearer when the issue is the influence of gender as discussed below.)

Though the differences in prediction of the Lederberg-Zuckerman-Fisher version of the rationalist hypothesis and the “garbage can” model may not be large, it is important to emphasize the differences in viewpoint. Rationalists see the organization influence as either supportive of researcher ideas or adversely impacting the possibility of those ideas being translated into studies. The organization itself is not a source of research ideas and does not socialize its researchers to suggest ideas for research relevant to organization needs. In the “garbage can” model, organizations are a source of ideas and/or do socialize their researchers to offer relevant ideas.

5. Gender

In the discussion of autonomy I suggested that gender differences in problem choice criteria greater among relatively autonomous scholars than among less autonomous scholars could be a crucial test of the rationalist hypothesis. But if this finding casts doubt on the rationalist hypothesis as a suitable foundation for an empirical theory of problem choice, does it imply that any current model is superior?

I believe, by default, any finding that casts doubt on the correctness of the rationalist hypothesis strengthens the case that a constructivist model is superior. However, a finding of *patterned differences* of any sort also throws into doubt the utility of a social constructivist perspective because it says that the research process is chaotic and that problems originate locally. This leaves the "garbage can" model as the best model (the resource dependence model only addresses the single factor of financial influence and makes no claims of gender differences in problem choice criteria.)

The "garbage can" model makes no assertions about patterned differences one way or the other. The model only says that problem choices are linkages occurring when independent or "loosely coupled" streams of demands, problems and constraints flow together. Gender patterned differences may suggest the kinds of problems researchers demand, if they can mobilize resources to create real demand. However, that is all the model says about what gender differences might do.

In short, whereas the rationalist hypothesis and the social constructivist model rise or fall depending on what, if any, gender patterns are found in problem choice criteria, the "garbage can" model neither benefits nor suffers from findings of a gender difference.

Section Two: Hypotheses of Problem Choice

It is now time to formally state hypotheses. However, before this can be done operational definitions are needed for many of the concepts. I will commence with "autonomy" followed by "theoretical vs. local origination" *etc.*

1. Definition of Autonomy

The rationalist position on autonomy is, testable in principle. The question that needs to be addressed is two-fold: (a) Are there differences in autonomy in problem choice between men and women scientists? and (b) Can these differences be shown to be predictive of differences in problems chosen?

If a valid answer is obtained to the above question , *i.e.,* if a valid test of the rationalist hypothesis is feasible, this could go a long way towards ascertaining if the rationalist hypothesis is a suitable basis for an empirical theory of problem choice.

The next issue that must be dealt with is how will "autonomy" be operationalized in this study?

Measuring "autonomy" is not simple. I decided to use a number of indicators, including an index of several indicators, to address this concept:

(a) Sponsor came to me and told me to do the research ("Yes" answer would be indicator of "low autonomy")

(b) I chose this problem because renowned researchers recommended research on it ("Yes" answer would be an indicator of "high autonomy")

(c) In choosing problem I studied, I first compared it with others ("Yes" answer would indicate "high autonomy")

(d) I have a great deal of freedom to choose the problems I want to study. ("Yes" answer would indicate "high autonomy")

(e) An index of the above four individual indicators. (see Chapter Three for details of the index construction).

2. Hypotheses of Problem Choice: Autonomy

In general, using these various indicators, I will try to ascertain the following:

(a) Are scientists generally "free to choose" the problems they study?

(b) Are there "gender based" or other *patterned* differences in degree of autonomy that scientists enjoy?

(c) If level of autonomy is controlled, do gender based differences disappear or will there still be gender based differences in problem choice?

Answers to these questions may bear on the degree to which we feel confident that the rationalist model may be a suitable basis on which to build an empirically based theory of problem choice.

If the data shows that women are less autonomous than men scientists, this would suggest that women indeed face barriers in problem choice compared to men scientists. Such a finding would be quite a satisfactory one to rationalist scholars provided that most men report that they are autonomous (*i.e.,* need only consult with fellow disciplinarians about what research to do and not have their problem choices set by nonscientists).

However, if the data shows that men also are not likely to be autonomous in problem choice, even if women are less autonomous than men, then the rationalist model is vulnerable to the criticism that it is a theory about a hypothetical (utopian) world.

In sum, the proportion of all scientists who claim to be autonomous, as well as gender differences in autonomy, will be important data for determining if the rationalist model or an alternative is a suitable basis for a model of problem choice.

Hyp. 3.1: **The proportion of scientists who regard themselves as having great freedom to choose their problems is greater than ninety (ninety-five) percent of the total.**

Hyp. 3.1a: **The proportion of scientists who believe they have great freedom to choose is positively related to being male.**

3. Definition of Theoretical Derivation

The definition of "theoretical derivation" is fairly straightforward. Generally, one thinks of this as a logical derivation from a general theory. As will be seen below, it is not actually so simple to operationalize the concept. I needed to try several different indicators that might be relevant, *e.g.,* a study recommended by renowned theorists, a study in which the investigator first compared two or more possible ideas for research before selecting one of them etc..

4. Hypothesis of Problem Choice: Theory Derived

I offer one hypothesis for this section:

Hyp. 3.2: **Scientist respondents will not always report either that problems are locally derived or theory derived.**

This hypothesis was tested using a "goodness of fit" chi-square test. As the technical appendix in Chapter Five shows, I used both a 90 percent and 95 percent criterion to stand for "almost always theory derived" (or "almost always locally derived").

I have a few observations about the indicators used to test the hypothesis.

At first glance, it would seem that measuring the concepts of "locally derived" and "theoretically derived" is fairly straightforward. Certainly, a number of indicators for theoretical derivation sprint to mind:

- Explicit statement by the respondent that the problem came from theory
- Denying that the problem was formulated by a nonscientist "sponsor"
- Explicit statement by the researcher that a renowned researcher suggested the problem as worthwhile

However, as the discussion of this issue in Chapter Five makes clear, testing the predictions of rival hypotheses for patterns of problem derivation proved more complex than I had expected. And the results were not as clear cut as I would have liked, although, I believe that my interpretation of them was the appropriate one.

5. Definition of Competition and Conflict

Evidence from the literature shows that competition and conflict are crucial concepts in the social sciences. Yet, there is a good deal of fuzziness in the concepts. Even so, it is important from the standpoint of the concerns of this study to see what guidance the literature offers on the relationship between competition (and/or conflict) on one hand and problem choice in science on the other.

It is important to keep two caveats in mind:

(a) The literature on problem choice and competition (or conflict) suffers from the same problems of lack of clear or uniform definitions of terms as the more voluminous literature on competition (or conflict) in other spheres.

(b) There is (as a result?) a lack of quantitative data on the relationship of competition (or conflict) and problem choice.

A. Benefits and Drawbacks of Competition

In at least one model of the development of science, competition plays a necessary positive role, according to Callon (1995:41). He explains that scientists work in those areas of research where the anticipated symbolic profits are likely to be highest (because the problems being tackled are considered important), and where there are still many areas of ignorance. The competition of other scientists, who are also the judges of one's own contributions, helps ensure that professional quality work is done.

Although the consequences of competition, especially interpersonal competition, are oftentimes positive as Callon implies, they need not be. R. K. Merton (1973:312; originally 1957) in his influential essay "Priorities in Scientific Discovery," observed that competition in the "realm of science, intensified by the great emphasis on original and significant discoveries, may occasionally generate incentives for eclipsing rivals by illicit or dubious means." Usually, Merton emphasized, the "dubious means" are "libel" or "slander" and even that is not especially common in his opinion.

B. Multiple Meanings of Competition

Competition is a term with several meanings. Unlike economists, who in price theory distinguish "perfect competition" from "imperfect competition" to characterize the effect of seller actions on the price of a good in different situations, social scientists do not give a consistent definition to the term competition. Caplow (1964:318), noting this lack of a commonly accepted definition of competition and conflict, suggests that it may be because conflict is a "very complex phenomenon," as well as "the central problem of society."

My review of the literature suggests that social scientists of science seem to use the term "competition" in two senses -- or to refer to two different phenomena:

(1) "head-to-head" or "interpersonal" rivalry between two or more actors trying to solve the same scientific problem first, and

(2) "structural" competition of scientists competing for honors, such as the Nobel Prize, appointments to prestigious positions, *etc.*, but not necessarily working on the same or similar problems as their potential competitors.

Mulkay (1991:10-11) seems, for example, to use these two senses of the term interchangeably. Sisela Bok, a philosopher, also makes no attempt to distinguish the different senses of the term. Thus, she wrote (1982:157):

> Modern scientists work under conditions of heightened competition as compared with their predecessors. On any one frontier of research, many scientists will be at work on very similar projects, often competing for limited resources. The level of specialization is such, moreover, that researchers cannot easily shift gears and turn to a different field. Jerry Gaston (1973:74), in a study of competition in science compares it to "a race between runners on the same track and over the same distance at the same time."

Zuckerman (1977) seems to have in mind interpersonal competition in discussing a Nobel laureate's desire to do an experiment before a friendly rival did it. She writes (1977:204) that a physicist indicated that in his own prize winning research:

> There were other people who were close to it, who were trying to understand this group of facts. One of them gave a paper at the annual meeting of the Physical Society after [we] had gotten [evidence of] the effect. ...Later on, I ran into [the scientist who presented the paper].... He wanted to talk about his experiment.... Finally, he said, "you know, I think that if one did it this way and measured the potentials - maybe from that experiment one could understand it." And I said "yes, I think that would be a good experiment."

It is clear that, in the Nobelist's own view, his colleague (who also is a friend as well as a friendly competitor) would have solved the problem in the not too distant future. The Nobelist could not afford to delay publishing, since he would lose "priority" for the work. James Watson's *The Double Helix* (1968), of course, boasts of the author's desire to be at Pauling, already a Nobelist, in the race to decipher the structure of deoxyribonucleic acid (DNA), the building block of most life. Merton (1973:287-289) adduces other examples of the desire by scientists for recognition of their intellectual accomplishments by their peers and by society.

Other writers on competition in science seem to have in mind structural competition including the "imperfect" competition of economists. Thus, Hilgartner (1995:312), discussing the specialty of molecular biology, states that it is "competitive" and that "tensions surrounding access to clones have been common." He observes that "access to biological materials often conveys a decisive competitive edge." While the rules of some journals in the field of molecular biology and human genetics require that "clones used be provided (post-publication) to any scientist who requests them," he notes that "waiting until publication introduces a significant delay. In human genetics, researchers often complain that clones they have been promised never arrive in the mail." And, he adds, "in some labs, there have been bitter disputes about the disposition of clones following the departure of personnel."

Zuckerman (1988:556), may have multiple types of competition in mind when, in a review of the status of the sociology of science, she states that "Knorr-Cetina . . . would attribute multiple discoveries in science to constraints imposed not by nature but by socially patterned understandings, processes of communication, competition and shared scientific culture."

Zuckerman emphasizes that she has no fundamental disagreement with Knorr-Cetina's analysis, only with Knorr-Cetina's insistence on how her views are at odds with Zuckerman's view "which attributes multiple discoveries to a shared knowledge base and a shared research agenda within the acutely competitive social framework of science." Elsewhere in the same essay, Zuckerman (1988:522) finds that "the great premium science puts on originality and peer recognition produces 'intense competition'."

C. Limited Evidence In The Literature For Existence of Interpersonal Competition

Though it might be gathered from the assertions of Bok (1982), Gaston (1973), Zuckerman (1977) and others that competition at the interpersonal level is common, it is not observed in many empirical studies where it might be expected to be found. Zeldenrust (1990:290-291), for example, insists that, while in his study, ". . . a striking degree of problem specialization and demand differentiation was found: none of the groups . . . was in substantive competition with any of the others."

Nevertheless, he wonders if, in the United States, as compared to the Netherlands, "conditions for mutual awareness and accommodation [between research teams] would appear to be less favorable, and substantive competition more aggressive."

Zuckerman herself observes (1988:538) ". . . in many documented cases, scientists prefer to avoid head-to-head competition with large numbers of other investigators and choose quieter, if less heady, domains of work."

In some cases where sociologists believe that they found competition in science, their study population (vehemently) disputed the finding. Mulkay (1991:10-11; originally 1974) remarks that when in a study of radio astronomers in the United Kingdom, he and his co-author

> . . . stressed the prevalence of competition among radio astronomy groups rather than cooperation... A number of our interviewees, and especially the group leaders, . . . stated that the two large groups in Britain had adopted an explicit policy of avoiding research overlap and that this policy... significantly reduced the likelihood of outright competition.

Mulkay further notes that his respondents had pointed to "much technical cooperation between the two British [radio astronomy] groups which would not be reflected in joint papers [between members of one group and the other]."

Though noting the objections of his respondents, Mulkay does not seem to believe that they might be correct. He points to the self interest of his respondents (as if self interest and truth inevitably conflict) as the reason for the "force and urgency" of the criticism he received:

> . . . when we stressed that radio astronomers were to a noticeable extent concerned with priority and professional repute, . . . that they led to competition and prevented active collaboration, we were seen by some participants to be undermining their whole professional life.

Mulkay, while modifying his position about an absence of cooperation between the British groups, indicates that, in his final report, he stuck to his main point about considerable interpersonal competition in the radio astronomy field.

Is Mulkay correct? Is there a great deal of interpersonal competition in science (or at least in radio astronomy)? There is evidence to the contrary in the literature -- aside from the vehement denials of his respondents that this competition existed.

Hagstrom (1986:47; originally 1965) provides an interesting argument against interpersonal competition lasting more than briefly on the basis of Darwin's observations of "competitive behavior in the animal world." Hagstrom says, "those who discover important problems upon which few others are engaged are less likely to be anticipated and more likely to be rewarded with recognition. Thus, scientists tend to disperse themselves. . . ."

This dispersal, he implies, is to avoid the struggle that working on the same problem as others is likely to engender. He quotes Darwin's observation (1958:82), ". . . the struggle will almost invariably be most severe between the individuals of the same species, for they frequent the same districts, require the same food, and are exposed to the same danger."

I agree with Hagstrom's point. Interpersonal competition of the sort I think Mulkay, Gaston and Bok have in mind is expensive. It requires twice the societal resources to investigate a problem when two or more groups in a society are both investigating it. It seems unlikely to me that the British or American governments desire that expense -- and they are the ones who pay for most research in their respective societies. I can believe that they require that researchers compete for the funds to do specific studies -- a situation of structural competition commonly occurring in my own field of criminal justice studies -- but I cannot believe that routinely they award multiple grants for the same research. Perhaps, in some rare cases, at least for a while, they fund different approaches to solving a problem, if the problem is truly significant scientifically and politically. This could lead to interpersonal competition between key personnel of different teams. (Funding two or more approaches to solving the same problem is an extraordinary step by grants-giving agencies; I am sure it occurs only in problems of utmost political importance and only until one approach emerges as more fruitful [*e.g.*, the "oncogene" model in cancer research]. Then, as Fujimura [1987] showed, one sees a "bandwagon" and again interpersonal competition dissipates.)

Though I doubt that interpersonal competition is common between scientists in the same country, it may occur more often between scientists from different countries. (Hagstrom's point from Darwinian biology permits this.) Mulkay's respondents, for example, agreed with him that (as research teams) they were in competition with foreign (*i.e.*, non-British) astronomy groups (1991:11). Such competition is probable, in fact, in issues of great national concern such as the race to build the first atom bomb or to discover the virus that is responsible for human immuno-deficiency disease.

Overall, however, the evidence reviewed here suggests that interpersonal competition is not common and thus not a likely source of problem choice. Structural competition, however, created by the scarcity of resources for research and the consequent need researchers have to attract resources others are also striving to get, is indisputably a significant factor in problem choice. Obviously, one has to adapt one's research in varying degrees to the requirements of funding sources or else one is unlikely to obtain resources required to proceed. However, I do not concern myself with structural competition in this study because (1) I believe that Shenhav (1985) has adequately covered this issue in his study of resource dependence's effects on problem choice and (2) neither the "garbage can" model nor the rationalist model dispute that there is resource dependence by researchers and competition for those resources. I do want to say, however, that structural competition can cause interesting scientist behavior such as the desire to avoid working on certain problems or problem areas (a point made by Zuckerman [1988]) probably because the scientist as an entrepreneur senses that the area is saturated with well established and talented researchers.

Before leaving the issue of competition in science, I also wish to note some observations of Merton.

Merton (1973:331, originally 1968) suggests that, in studying competition as a source of problems, the discipline, or at least specialty, may matter. He hypothesizes "hot fields . . . tend, at least for a time, to attract larger proportions of talented scientists who have an eye for the jugular concerned to work on highly consequential problems rather than ones of less import." In other words, Merton *specifies* the relationship of competition and problem choice. He further hypothesizes (*loc. cit.*) "that levels of interaction between workers in the hot field, particularly at the leading edge, are unusually high." And he expects (*loc. cit.*) "that kinds and degrees of competition differ among specialties ...," suggesting that the "hot field" may be a specialty and not an entire discipline.

The lack of empirical evidence in the literature about interpersonal competition of which I spoke before digressing to discuss Merton's observations may simply be a result of using unreliable means to detect interpersonal competition. For example, Zeldenrust relies on archival data and "interviews with informants." However, the archival record may not be a valid means to detect competition. Quite possibly, the sense of wanting to do something (or almost needing to do it to meet some psychological need) before a rival overseas or elsewhere in Europe does it is not necessary to record in the documents prepared by research teams to meet requirements of Dutch law. Merton sagely notes (1973:326, originally publ. 1968): "... the etiquette that governs the writing of scientific papers. This etiquette, as we know, requires them to be works of vast expurgation stripping the complex events and behaviors that culminated in the report of everything except their cognitive substance."

The etiquette of which Merton speaks may equally apply to the documents that Zeldenrust examined, leading him to miss the keen desire for priority against adversaries known or unknown that may drive many of the people he studied.

Quantitative data on conflict as a factor also are thin or non-existent, perhaps because in that case also, there is a problem in validly measuring "conflict." Finding even anecdotal evidence of conflict between scientists is quite difficult. Newton's efforts to diminish the importance of Leibniz's contribution to the development of calculus may be an example of extreme conflict. But the conflict seems to have occurred *after* work on the calculus was completed and, thus, one could not say that, in this instance, conflict, like necessity, was "the mother of invention."

D. The Role of Competition and Conflict in Problem Choice: My Perspective

If I understand the rationalist position correctly, it argues for *two-way* causation between problem choice and competition. First, it maintains that since scientists try to solve the most important problems that they can, an "intensely competitive social framework" (as Zuckerman puts it) is created as many scientists more or less simultaneously select the most significant issues in their discipline and try to win priority by solving them first. Cozzens (1986) illustrates this situation nicely in her study of rival drug research teams.

Second, Zuckerman argues that many scientists, knowing the intense competition surrounding research on certain issues, shy away from those very important questions and choose "less heady" domains. What is crucial to keep in mind is that the rationalists predict a *correlation between* problem choice and interpersonal competition. The constructivist models, including my own "neo-constructivist" perspective, do not predict such a correlation.

However, even if the rationalists are correct in predicting a correlation between interpersonal competition and problem choice, I believed that it would more likely be found in fields where researchers often work alone or in small groups and that are easy to enter. I was much influenced in my thinking by an observation of Merton [1973:331]:

> . . . scientists often do not know who else is engaged in similar work, and this lack of information generates its own brand of pressures and anxieties. Competition is less often experienced as a negative, though often intense, personal rivalry; it tends to become a diffuse pressure to publish quickly in order not to be preempted by unknown others.

It seemed to me that in the situation described by Merton, the scientists are subjected to *cross pressures*. On one hand, there is the pressure to publish so that rewards of research funds, promotions, *etc.*, are gotten. On the other, there are the norms of science requiring a certain disinterestedness, a willingness to share findings for the greater good. I believe these norms are *stronger*, or more strongly internalized, however, when the scientists' competitors are colleagues and friends, rather than anonymous "somebody's out there." Faced only with competition from unknown others for research funds, I expected that the scientists are more likely to exhibit competitive behavior that evades, if not actually violates, the norms of science.

I further reasoned that the competitive pressures in the social sciences and some areas of biology might be greater than in the hard sciences (*e.g.*, physics, chemistry) because of lower entry costs to practicing in the social sciences or biology. I reasoned this way because, in my experience, I have often seen mathematicians and other mathematically competent scientists and engineers working on social science problems, whereas I seldom, if ever, saw social scientists tackling problems in mathematics, physics, *etc.*

These mathematically trained scientists, moving back and forth between hard science work and social science (and perhaps back again?), I think of as "generalists" in the sense meant by Freeman and Hannan (1983), as presented in Aldrich and Marsden (1988). The generalists are able to outperform the specialists in certain kinds of environments according to Freeman and Hannan; the specialists outperform the generalists in others. Specifically, Freeman and Hannan:

(1) maintain that in stable environments, including those "changing in predictable ways," the specialist strategy " is always an optimal one" because "specialized forms can

concentrate on that portion of the environment in which they do best."

(2) Given "uncertainty" in the environment, "the situation is more complex." The key to determining the optimum strategy is knowing "the extent to which organizations can tolerate extreme environmental conditions." As summarized by Aldrich and Marsden, Hannan and Freeman contend that:

> If many organizations can survive at the extremes, generalists would be favored in all uncertain environments. If, however, most organizations cannot survive at any extreme, specialists would be favored when change is rapid ("fine-grained") because they will encounter favorable conditions often enough to ride out difficult environmental states. No form would do very well when change is slow and unpredictable, though Freeman and Hannan hypothesized that generalists would outperform specialists.

6. Hypotheses of Problem Choice: Competition and Conflict

Given this line of reasoning, I wondered if:

(1) my respondents who were social scientists (and other researchers in the "inexact sciences") would be more likely to experience competition and conflict than respondents in the exact sciences;[7] and

(2) my respondents who were doing research in the inexact sciences, since they probably also worked in areas that were related to public controversies (*e.g.*, crime control strategies, reproductive freedom), *might* themselves do studies, one of whose primary aims would be to discredit the ideas of rivals. Or they might know of others in the field who chose problems for this reason.

More formally, I hypothesized:

Hyp. 3.3: **Competition between scientists in the inexact sciences will be more intense generally than that in the exact sciences.**

Specifically, I evaluated the following relationship:

Hyp.3.3.1A: **Researchers in the exact sciences (*i.e.*, social sciences, biological/medical sciences) are more likely to report that one of the reasons they chose their most recent problem was a desire to do the research before a rival did it.**

I also evaluated the following relationship:

Hyp. 3.3.1B: **Researchers in the inexact sciences (where problem choice is often influenced by a desire to contribute to public controversies) will report that "some scientists do studies to discredit their rivals" or to "support the positions of their friends" more than researchers in the exact sciences. (compare Merton 1973: 330-331)**

Differences between disciplines in regard to the role of conflict and competition in problem choice were not the only patterned differences I believed might be important. I expected that women scientists more than men scientists might notice, or suspect, conflict or competition underlying problem choice of other scientists.[8]

Therefore, in formal terms, I propounded the following:

Hyp. 3.4: **Women more than men will report that "researchers sometimes (or more often) do studies to embarrass their rivals" or "support their friends views."**

Hyp. 3.5: **Men more than women will report "a desire to do something before someone else does it was a factor in their problem choice."**

Finally, reasoning analogously to Sulloway (1996:7) that women will emphasize different criteria than men in problem choice, as part of a strategy to differentiate themselves from men advantageously and thereby encourage investment in their scientific work I investigated the following hypotheses:

Hyp. 3.5A: **Women will follow different problem choice strategies from men to differentiate themselves from men and gain investment in their research.**

Hyp. 3.5A.1: **More women than men will assert that they take into account if the problem is related to current political/social concerns.**

Hyp. 3.5A.2: **More women than men will assert that they take into account if the problem can be done within applicable time constraints before they choose to work on it; and**

Hyp. 3.5A.3: **More women than men will assert that they take into account whether a renowned member of the discipline has recommended research on the problem**

2. Definition of Organization Culture

Organizations shape the perception of their members, sometimes quite dramatically. Festinger (1957:244 passim) provides several examples of how, even in the face of compelling contrary information, members of certain religious organizations ("doomsday cults") held to beliefs to which they were committed. Festinger attributes this persistence of the incorrect belief to social support those holding the belief gave to one another.

Kuhn's (1962;1970) description of how scientists often tenaciously cling to old paradigms rather than accept new ones while others not acquainted with the old accept the newer paradigm seems to suggest that scientific theories arouse the same intensity of psychological support in their adherents as the "doomsday" predictions in Festinger's examples. Consequently, those holding to the older paradigm may experience dissonance in the face of new contrary information. And these "holdouts" in science may continue to support one another in their beliefs about the prior paradigm.

If organizations can indeed shape the beliefs of their members, it seems plausible that the organization accomplish this in either (or both) of two ways: (1) they foster an appreciation of practical (i.e., non-theoretical) research among their employees' or (2) they foster a belief in the greater importance or correctness of different paradigms from those the scientist may have been committed to when she was hired.

The change in scientist commitments occurs during the socialization of the scientist into the organization. This socialization begins, perhaps even before the scientist joins ("anticipatory" socialization) and continues throughout her career.

The researcher commitments, to theories and to various criteria for assessing the worth of particular problems, form essential constituents of the "organization culture" of the research team(s) in which science work is performed. This organization culture is important to understanding the problem characteristics of problems team scientists offer either in response to demands from the organization or to probe organization demand.

The organization culture's effect on problems etc., is not to be confused with the organization culture's "aim," which, at the least, I maintain, is to influence the manner in which the unit seeks to control its research process in the face of external demands. (More generally, as Zeldenrust emphasizes [1990:29], research organizations try to maintain control over the "parameters and the configuration of the research process, that is, over the contents of each of the flows [demand, resources, problems and constraints] and over their interrelationships respectively.")

Zeldenrust devotes considerable attention to elucidating the control strategies of the teams of biotechnology researchers that he studied. However, he never explicitly considers the role of "organization culture" (the term, for instance, does not appear in his text!) even though he hints at its possible significance in constructing the team control strategies in case studies he presents. Perhaps the clearest example of the possible importance of organization culture is his case study of "the dialectics of composite materials" (1990:175-185). Zeldenrust notes that the research team studying polymer mixtures "throughout its twenty year history" used a "single research approach: model and theory oriented product research." Despite ridicule in the early years from industry colleagues about his research topic ('How are your little mixtures doing?' they jeered),[9] the university based principal investigator on the team tenaciously persisted in his approach until it paid off. Zeldenrust also observes that the group decided to "move from mixtures of polymers to 'reinforced polymers' (composite of polymers and non-polymers) after it had gained complete control of its raw material and could routinize variations. Furthermore, it abjured continuing to study simple polymer mixtures - a route that "could have led to 'easy publications'."

This example illustrates the possible value of organization culture in insulating the researchers from the possible loss of self-esteem when undertaking a risky research program not well appreciated at the outset by professional colleagues. It also shows that the organization culture, (which emphasized commitment to tackling the hardest problems the group could solve rather than easy work that might result in higher productivity scores, *i.e.*, more papers published per capita but less significant intellectually) probably propelled the team to continue to seek challenging work that would allow it to utilize the theoretical/model building strategy that had already yielded significant intellectual dividends.

Although illustrating the possible value of organization culture in a "garbage can" model of problem choice is desirable, it is hardly sufficient.

Before I can utilize the concept of organization culture in my study I need (a) to define the concept, and (b) to develop a method to demonstrate how it contributes to problem choice. I follow Sackmann (1991:34) who defines the organization culture as:

> . . . sets of commonly held cognitions that are held with some emotional investment and integrated into a logical system or cognitive map that contains cognitions about descriptions, operations, prescriptions and causes. They are habitually used and influence perception, thinking, feeling and acting.

Sackmann says that the cognitions referred to in her definition become commonly held in processes of social interaction. Furthermore, she says (*loc. cit.*)

> They [cognitions] can be introduced into the organization based on outside experiences, they can emerge from growing experiences, they can be invented and/or negotiated. In repeated applications they *become attached with emotions and assigned with degrees of importance* - also commonly held. They are *relatively stable over time* and accumulated in the form of different kinds of cultural knowledge (emphasis original).

As it stands, the concept is stated too generally, and I will need to use a lower level of abstraction than Sackmann if I am to work with the term. Below, I state my hypothesis and indicators of concepts, I shall describe how I operationalized this variable.

It is important here to note that organization culture is a work-related culture. It develops by (a) importing resources of various kinds from the larger environment beyond the identifiable team, itself an environment for its members as White [1992] would emphasize, and (b) originating group values and developing other resources internally through the face-to-face interaction of group members.

The organization culture is the basis for constructing control strategies that the research unit will use to perform its work; the control strategies in turn, determine, for example, what problems it will seek to work on among those identified in the external environment as problems whose solution there is a willingness to fund; what kinds of funding sources for locally developed problems will be sought; what scientific audiences, as indicated by journals in which unit member articles are published, are relevant; *etc.*

The research team uses its control strategies primarily for *accumulation* (*e.g.*, of problems the unit wants to address, of funds it wants to obtain, personnel it wants to add to address certain exigencies, *etc.*). However, there are theoretically important exceptions to the general aim of accumulation, a principal one being the "bandwagon" (see Fujimura [1986], also quoted in Zeldenrust [1990]) where problems suggested by one model that the group has acquired resources to study must be dropped because of a paradigmatic change in the field. Such paradigmatic change makes problems derived from the replaced model pointless or unproductive to pursue. While this kind of change is rare in most disciplines ('normal' science), in certain 'hot' fields such paradigmatic change is an important practical issue.

The control strategies of teams doing basic intra-disciplinary research is fundamentally different from that of units or teams doing multidisciplinary research because the organization cultures are different. Thompson (1967:136) suggests why this should be expected theoretically:

- The more numerous the areas in which the organization (*i.e.*, research unit) must rely on the judgmental decision strategy, the larger the dominant coalition.

- The less perfect the core technology the more likely it will be represented in the dominant coalition.

- The more heterogeneous the task environment, the larger the number of task environment specialists in the dominant coalition.

Thompson indicates (*loc. cit.*) that "under these propositions, we would expect to find the small dominant coalition, perhaps the single individual, in organizations repeating standardized technological activities to produce standardized results for standardized customers or clients."

Of course, research as it is understood here (*i.e.,* research at the 'frontier') does not have a perfect core technology, and problems at the frontier tackled by scientists require judgment about diverse issues, too diverse usually for a single scientist to be expert in all. Therefore, the dominant coalition is larger, as suggested by collegial structure research units. Shenhav adduces survey data (1985:130) that corroborate Thompson's prediction. Type of work is much more relevant to work structure than is whether the team is externally controlled or internally controlled. Shenhav (1985:130) suggests why this is true. "We would expect that externally controlled teams are interdisciplinary in nature," he says, "conducting more applied and developmental types of work. These types require a higher degree of interdependence and interactions than basic research teams." Therefore," he adds, "it is possible ... that the type of work in externally controlled teams requires an internal collegial structure."

Organization culture through its impact on control strategy plays the role in 'negotiation' of problems the unit will address that Fujimura (1987) had assigned in her perspective to the individual researcher. It is the unit that is actor in the priority disputes (see Cozzens [1989] quoted in White [1992]) because the unit is seeking resources (see Busch, Lacy and Sachs [1983] discussed above) to continue as a 'going concern' in the research business.[10]

Once the concepts of 'organizational culture' and competition/conflict are included and their role in problem choice clarified, the Zeldenrust model has the advantage of both the rationalist and the constructivist models. This is clear when comparing both models on how they address competition and conflict as a source of problem choice.

The rationalist model directs attention to competition/conflict in problem choice because it is presumed that the stock of identified problems is finite, and the really significant ones are a still smaller subset of all problems. Thus, especially in hot fields attracting the attention of scholars with an instinct for the really crucial issues, there should be competition and perhaps conflict as researchers vie for the honor of solving the major issues and obtaining recognition for their priority.

What role does competition play in problem choice according to the social constructivists? Basically, the social constructivists focus on unique conditions in laboratories and their collective influence on problem choices. However, since the social constructivists acknowledge external demand as relevant, they allow at least an *indirect* effect of structural competition. That is, competition between embedding organizations might indirectly influence problem choice of the research team by affecting demand.

The "garbage can" model's position is similar to the social constructivist perspectives. However, by including an organization culture variate in the "garbage can" model, I allow structural competition to influence not only embedding organization demand but also demand from the organizational set of the research team. (This issue of organizational set influence will be explored in Chapter Five.) Here I just wish to point out that in my opinion, without the organization culture variable added to the model, the research team in the Zeldenrust model seems somewhat like a jellyfish floating along in an ocean current testing its immediate environment for problems or demands or resources but not actively swimming toward some problems, *etc.,* while ignoring other less interesting problems, *etc*. However, having the organization culture direct the unit control strategy, which, in turn, pushes the unit to seek specific problems, specific resources for their solution and recognition (preparing articles for publication in professional journals) of their priority, fixes a significant defect in the Zeldenrust model.

At the same time, by adding "organization culture" to it, the Zeldenrust "garbage can" model has the advantage of simultaneously being able to explain how a locally developed problem or how a theory derived problem is chosen for research. Now it is possible to say the choice depends on the organization culture which weighs items present more or less heavily depending on which seems most relevant or important in preserving or enhancing the group's well being.

Recall that, in the constructivist tradition, the Zeldenrust model makes no assumption about the source of the problem. His model only requires that there be a demand, adequate resources, and a problem before a linkage, *i.e.*, a choice, is made. Constraints may be present also (the model allows for them, but their presence is not always required). In fact, Zeldenrust generally leans in the direction of local origination perhaps because the Dutch biotechnology teams generated most of their problems locally in the course of research they were doing rather than by turning to the professional literature or to informal communication with 'invisible college' members (Zeldenrust 1990:308).

Earlier in this chapter I indicated that the definition of 'organization culture' provided by Sackmann (1991) was too general to use as it stood and that I would provide an operational definition. I shall do that here.

For my purposes, organization culture is a three-dimension variable whose dimensions are:

(a) Cosmopolitan/Local

(b) Profit-oriented/Professional

(c) Mathematical/Eclectic

These variables are dichotomous in my definition and eight (2^3)combinations are possible logically. I will define each dimension in turn before considering how I dealt with the overall variable of organization culture.

Cosmopolitan/Local

The terms "Cosmopolitan/Local" are due to Merton (1957). In this essay I use them to describe *organizations* from the standpoint of whether they encourage researchers to discuss possible topics for research with researchers from many disciplines and many organizations, if need be; or discourage researchers from discussing possible topics with the widest possible groups of colleagues within and outside the organization. The former I label "cosmopolitan" organizations and the latter I label "local." The data for this variable are the subjective opinions of the researchers within the organization, not the official positions of the organization. In some sense I regard the subjective perceptions of the researchers as more valid, as more reflective of the "way things really are" but I have not attempted to compare the views of my respondents against official positions.

Profit-oriented/Professional[11]

"Profit-oriented/Professional" organization refers to the fact that the organization is *legally* (*i.e.,* officially) oriented towards making a profit or not officially oriented (*i.e.*, professional). Note that "professional" does not mean that the organization has no concern with obtaining a net surplus from its activities. It only refers to the fact that the organization is officially not primarily concerned with making a profit which the shareholders would receive (a part of) in the form of dividends.

Though I essentially depended on my respondents to indicate what the status of the employer is, I may on occasion have taken the liberty of changing a response that I knew to be incorrect, *e.g.*, a university is not a "for profit" organization though it may engage in research, *etc.*, to bring in revenues above its costs.

Mathematical/Eclectic

Mathematical/Eclectic refers to a commitment to a particular approach, a highly technological/mathematical approach. Problems suitable for handling by means of such a distinctively mathematical approach are chosen in preference to others that are not suitable for mathematical/technological approaches (*e.g.*, qualitative research in the social sciences).

8. Hypotheses of Problem Choice: Organization Traits

I assert that the types of organization culture predict the criteria the research team uses to assess any research problems for study. More abstractly, the type of culture predicts the dominant flow that the organization moves close to the other flows to enable linkages to occur.

Studying entire organization cultures requires data on a large number of cases. My data set is not large enough to compare several organization cultures. However, it is possible to predict certain criteria of problem choice from particular organization characteristics, which is advantageous with my small set of cases. And some of these particular organization traits may be emblematic of entire cultures, *e.g.*, “cosmopolitanism.”

One type of flow – “problems” – is a strategic research site for testing the relationship of organization traits and criteria. Based on my hypothesis presented earlier that the organization influences scientists perception of what criteria are important for research, perhaps even of what paradigms are theoretically important, I argue that:

Hyp. 3.6: There should be a negative relationship between the criterion “the problem was suggested by theory in my own field” on one hand and the organization trait of “the organization wants me to find research problems by considering whether the work has commercial applications.”

Hyp. 3.7: **There should be a significant positive relationships between the criterion of "I found the problem I wanted to do by reading widely outside my own field" and the organization trait of "distributing the organization mission statement and/or other policy statements expressly encouraging MDR by staff."**

Hyp. 3.7a: **There should be a significant positive relationship between the criterion of "I found the problem I wanted to do by reading widely outside my field" and the organization trait "the work site encourages MDR projects by (a) specifically seeking out grants for MDR or seeking funds allocated by the work site for MDR work," and (b) "providing grants of funds specifically to help MDR projects."**

Hyp. 3.8: **There should be a positive relationship between the criterion "I considered whether the results of my work could be patented," and the organization trait "the organization wants me to consider whether the problem chosen can make use of equipment the employer wants to see utilized."**

A bit of commentary on each hypothesis is in order here. Hyp. 3.6 already begins to suggest how deeply the organization influences the values of the scientist. When she has completed her rigorous professional training, it might be supposed that she has learned to almost automatically assess problems by asking if they contribute to theory in her own field or are derivable from one of the paradigms of her field.

However, as the first hypothesis dramatically illustrates, scientists will not automatically appraise problems against such a criterion once they are members of the *work* organization. They will look at whether the problem makes theoretical sense or contributes to theory even in other disciplines. (This is not so difficult for them to do because scientists have learned fundamental ideas of many different sciences during the course of their formal education.) Their commitment to the main purposes of their employer, *i.e.,* profitable applications of research that contribute to the bottom line, will overcome the effects of their professional socialization in this regard.

Hyp. 3.7 argues that researchers will read widely outside the literature of their own discipline given that their work site encourages MDR both formally (policy documents) and substantively (grants for MDR). It is possible, of course, that the researchers are only looking for new areas to apply the paradigms of their own discipline. (Harrison White, focusing on social scientists in particular, warns that this kind of search can lead to "explainings away." He adds, "Explanations soothe. On a micro scale such studies are known as "cooling the mark out" with plausible ex post account." The dangerous consequence is that "a main outcome is to disperse responses rather than have them possibly cohere into joint projects." (White [1992a:92]) There is no reason to suppose the danger only occurs in social science research. However, the researchers may also be looking for theoretical ideas outside their own discipline to apply towards solving work organization problems. The process suggested in Hyp. 3.7 may occur simultaneously with the process indicated in Hyp. 3.6: commitment to solving problems suggested by theory in ones own discipline weakens as commitment to solving problems of concern to the work organization grows even when the latter means widening one's theoretical interests to those outside one's own field. Of course, it is also possible that commitment to solving problems in one's own field grows *along with* commitment to the organization, but , in principle, these two commitments compete for the scientist's loyalty.

Hyp. 3.8 already begins to show that the scientist, as a member of the work organization, has learned to be concerned with criteria other than the theoretical significance of the problem. This is just as important to many work organizations as that the scientist not look at problems exclusively through the narrow prism of her own-professional training – a rigorous form of socialization which instilled in her a deep appreciation for the importance of her disciplines' paradigms (and many related disciplines as well).

Clearly, the criterion in the second hypothesis of wanting to do research leading to a patent is evidence of the scientist's having embraced new criteria since joining the work organization consistent with the purposes of the work organization.

The organization trait in Hyp. 3.8 warrants some comment. Organizations wanting particular equipment utilized in research have at least implicitly endorsed certain theories because equipment represents a theory. For example, special instruments for "atom masking" or for investigating the structure of a molecule represent theory about the respective phenomena.

Of course, the organization's commitment to particular paradigms, as represented by the desire to see certain equipment utilized, may not be primarily an intellectual commitment. The desire to see equipment utilized can have other bases, *e.g.*, that research be done economically.

Hyp. 3.8a: **The criterion of "considering whether I can be covered by a colleague's patent" is positively related to "the organization wants researchers to consider whether the problem can make use of equipment the employer wants to see utilized."**

The foregoing hypotheses are only meant to suggest how deeply and how broadly the embedding organization influences the criteria the scientists who are principal investigators bring to bear in appraising problems for possible research. Other examples of criteria that come readily to mind are: the criterion of "political sensitivity of the problem to federal, state, or local governments"; "considering whether I could acquire special equipment I do not presently have" and "considering whether I would have to borrow research techniques the employer wants to see utilized." (see Chapter 4, p. 165 *passim*).

Up to this point, I have focused only on organizational effects on criteria. However, it is possible that some criteria are influenced *jointly* by organizational factors and personal factors of the scientist respondent. The criterion "Problem could be studied within applicable time constraints" appears to be an example. I shall not state a hypothesis about this; in Chapter 4 when I present my findings, I discuss this example and the issue of determining the relative effects of each of the predictors on a dependent variable.

9. Definition of Gender Differences[12]

In his monumental work *Born to Rebel*, Sulloway comments (1996:149) "that siblings face differing behavioral contexts in their different family niches . . . . Being female does not constitute the same experience for a first born as it does for a later born. The same conclusion is true about brothers . . . ."

Sulloway raises an interesting point. The first arrival in the family experiences a different family than the later born. And Sulloway follows with the observation that the second born chooses strategies to accentuate the differences in order to obtain greater parental investment in the second born. As Sulloway notes (1996:171), it is not just men who cause science to be dogmatic, impersonal and competitive. Science and letters have always been dominated by first born individuals who excel at traditional forms of learning.

As I read these lines, I wondered: Is it possible that by some analogous process latecomers to the university (a) experience the university differently than the first arrivals and (b) seek to accentuate these differences to cause some "investment" by others to help the latecomers fit in? Specifically, can we expect women (who are academic latecomers) to experience the university differently than men and also to accentuate the differences from men, for example, in how they do research (a primary faculty responsibility at the better schools)?

Let me state the issue another way. I note that since women are late arrivals in the ranks of academic scientists, there are far fewer senior women scientists, especially in engineering and other exact sciences than would be expected based on the proportion of women in the population. Secondly, women scientists will have generally had less collegial relationships than male students with their professors during their graduate training (*see* Fox [1995:220]). The result is that women will not have access to the most obviously promising topics for research nor to the resources of senior male scientists to the same degree as men. Therefore, I reasoned that generally women will have to adopt riskier strategies of problem choice. For example, more women than men may be (1) working in marginal specialties and interstitial areas between disciplines where competition with dominant males is less likely. Alternatively, (2), like Barbara McClintock, they may be seeking niches that are overlooked but have promise and tenaciously working on the problems in those niche areas despite ridicule, funding uncertainty, and other obstacles. The first strategy, the most common choice, has features such as: (a) avoid competition with senior male faculty (who have more research experience), (b) seek to work with experts from another field who can teach you (but who will not damage your claim to priority within your own field), and (c) do what you can without special equipment when necessary since this may be difficult to obtain for the less senior scientists.

The second strategy has some elements in common with the first. However, the main difference is that in the second strategy, the researcher seeks to become the *dominant* researcher in a niche of real importance. This may discourage male or female competitors from entering that specialty. (In the first strategy the women may be more likely to change specialties for one or another reason including not wanting to go up against strong male competitors attracted to a suddenly interesting specialty that the women were working in.)

The idea that problem choice may be gender related because of differential access to mentors apparently did not occur to the many writers on the subject of women in science because I could not find any such suggestion in the literature I reviewed. It seems that regardless of whether the writer is a resource dependence school thinker such as Shenhav (1985), a constructivist such as Knorr-Cetina, or a rationalist such as Zuckerman, the assumption is that problem choice processes are unrelated to gender differences.

It is possible, however, that the idea of gender differences in problem choice could have occurred to the neo-constructivist Zeldenrust if he had theorized along the following lines: Assume that, as Zeldenrust argues, the process of problem choice is non-rational and that multiple linkages are possible between problems and demands. Then, if Shenhav (1985:115) is correct that scientist prestige is relevant to bargaining ability in the acquisition of resources, perhaps it is not farfetched to imagine that among scientists gender is related to prestige (*i.e.,* women scientists enjoy less prestige overall than men as evidenced by salary differences, name recognition among colleagues, *etc.*). And if this is so, then even the resource dependence model should predict that women scientists perceive less autonomy in problem choice than men do. That is, women scientists generally should report that client demand, *i.e.*, nonscientific audience pressure, is more compelling in problem choice than it is for men.

Before proceeding further with my theoretical development on gender, I want to digress for a moment to point out that problem choice of women compared to men may be a "strategic research site" for understanding an issue in the sociology of science that is the subject of much dispute; I am referring to the "gender productivity difference" question.[13] Cole (1979) and Cole and Singer (1991) have argued that men are more productive scholars than women. They have suggested that this productivity difference helps explain why women in academia, especially at prestigious universities, are underrepresented in the ranks of full professors (1979; 1991, 1993). And with Singer (Cole and Singer [1991]) Cole has offered an hypothesis of "limited differences" that he believes may be a useful theoretical tool in explaining why women produce proportionately fewer scholarly papers than men.

Cole and Singer's model is a probabilistic model that is intuitively appealing. It posits that perhaps women begin their academic careers with subtle disadvantages such as a smaller pool of outstanding scholars willing to sponsor their dissertation research. Also, it suggests that perhaps women differentially feel more discouragement than men when rejected for tenure in a prestigious university, or for a grant for research. At the same time, women may be feeling less positively motivated than men by rewards such as granting of tenure in a prestigious university, etc. Over time all of this causes a cumulative disadvantage for women and that disadvantage adversely impacts their productivity.[14]

Cole (1979) suggests that studies of the determinants of the gender productivity gap might look at factors in the development of women *before* they reach graduate school. And the sophisticated model he and Singer (1991) propose is consistent with that suggestion since the limited differences they posit may begin before graduate school. However, even allowing that Cole and Singer's statistical model may be correct, I maintain that one should first look at the environment of the research centers where women scientists work.

Specifically, I would look at the process by which women go about making research problem choices within their embedding organizations to see how women scientists' productivity could be adversely impacted.[15] My reason for doing this is some recent observations of Mary Frank Fox (1995). Fox (1995:221) has pointed to a compelling reason to suppose that women would choose problems different from those of men -- resource differentials. Thus, she reports that Feldt (1986) had found that among "scientists appointed as assistant professors at the University of Michigan . . . women and men received different treatment, even at this early career stage. The men got more start-up monies for their labs, better physical facilities, and better placement within existing projects with funds and equipment." And, as Fox notes, "placement in projects" is relevant for collaboration. This is important, she continues, ". . . because solo research is difficult to initiate and sustain in science." Sometimes, the resource differential is subtle. For example, Fox observes that, "Women may have more difficulty finding and establishing collaborators and they may have fewer collaborators available to them." One study (Cameron, 1978; quoted in Fox [1995]) supports this notion, in reporting that men have significantly higher numbers of different collaborators. Further, Fox argues that Long's data (1992) show that in biochemistry, women are much more likely than men to collaborate with a spouse. To the extent that this collaboration of women occurs with a spouse instead of multiple others, these data also may suggest a more limited range of collaborators among women compared with men.

It is apparent in these observations that women face both subtle and obvious obstacles to pursuing research. For women especially, problems requiring heavy funding, problems requiring collaborations with specialists other than their spouses or close friends, may be foreclosed entirely. Even if in theory the women could work on any issue they wish, the competitive disadvantage adversely impacts women senior investigators far more than men, and it may be sufficient to keep even talented women from working on the most interesting problems they identify, or even on the problems they identify most quickly as worth studying.

Consider the possibility already raised that women scientists do not have the close collegial relationship with senior faculty members that male scientists do. As a result, the women must wait to check out the significance of a problem they are interested in studying by going through the professional literature whereas men may feel more comfortable with simply calling up a respected senior faculty member and asking what he thinks of the problem's significance.

Compared to simply checking verbally with a senior scientist, seeking out problems in the literature that renowned researchers consider important may be particularly time consuming. In the process, these

women may overlook problems "in their own backyard" that could be fruitful. Men, availing themselves of the possibility of talking to senior faculty, may be advised of worthwhile problems "in their own backyard" by the senior faculty member and/or confirmed in the hunch that the problem they brought to the attention of the senior faculty member is indeed important. The men's easier access to senior scholars may be reflected in the former's higher productivity. These possibilities are not discussed in the literature I have seen.

The reader may infer from my remarks that I think women scientists should simply learn how to talk to senior faculty and all the problems of male-female scientist productivity differences would vanish. Nothing could be further from the truth. The reason women do not have a close camaraderie with senior male faculty, especially in engineering and some other sciences, if Mary Frank Fox (1995) is correct, is that perhaps the men faculty discourage women from having such easy working relationships. This suggestion is implicit in Singer and Cole's "limited difference" model.[16]

In terms of the "garbage can" model, my point is that neither men nor women choose problems "rationally." But, women may be facing obstacles in the process of identifying problems that men do not generally face. And because of the obstacles (or whatever), the number of linkages (which is what I see as problem choice) they make is fewer than the number that men make. Though all researchers do considerable planning, ambiguity affects the choice process. You may have an interesting idea for research but you must first line up resources to pursue it. Will you be able to do this? And will you be able to do it before someone else with the same idea solves the problem? These are some of the uncertainties that make the process ambiguous. In the case of women, however, perhaps ambiguity affects more aspects of the problem finding process than for men.

(Alternatively someone is offering money to anyone who can solve a problem [as the former defines it]. Can you find a problem that addresses the issue they raise, that is, a problem whose solution meets their demand? This too, is not certain, *i.e.*, it too is an aspect of the "ambiguity" to which Zeldenrust devotes considerable attention. Again in this case women may face more uncertainty, or more areas of uncertainty, than men.)

These observations give a new meaning to the general argument of Zeldenrust. Recall that, in the most abstract sense, demands "*seek*" problems; problems "*seek*" demands; and resources "*seek*" to be used. A linkage *may occur* among the demands, problems, and resources (and constraints, if applicable). Yet ultimately the resources -- herself and other scientists -- may be put to work on a problem different from what the proposing party (the scientist) suggested, but one that is (at least, minimally) acceptable to both her and the sponsor.

Furthermore, recall that besides being non-rational, the process of problem choice is *nonrandom* in the Zeldenrust model. Partly, this is because as findings from earlier studies (*e.g.,* Shenhav [1985]; Nederhof and Rip [n.d.]) suggest scientists are not autonomous in problem choice. To paraphrase an old saw, "Problem choice in science is too important to be left to scientists." The organizations employing scientists participate in, even control to a considerable degree, the choices of problems to ensure that certain constraints on problem choice are considered (*e.g.,* that the problems chosen reflect the most important values of the ideology to which the employer adheres) (see *e.g.,* Westfall [1977]; Zeldenrust [1990]).

Now I am suggesting in light of research on gender differences in productivity that the operating dictum of scientists' employers may well be "Problem choice is too important to be left to scientists, especially if they are women." This, of course, is a view at odds with the social constructivist view and the rationalist view as I understand them.

In the social constructivist view patterned differences between the genders seem not to be even a possibility since any underlying regularities do not seem plausible. (They challenge the social constructivists' insistence that all problem choice are "chaotic.")

In the rationalist model, gender differences are also unexpected. Problem choices are based on inherent attractions/flows in the problems themselves. Zuckerman as a structuralist would not dispute that women face difficulties in scientific research that men do not face, but she would not imagine that a model of problem choice could address that issue. However, the Zeldenrust model *can* accommodate that matter nicely.

10. Hypotheses of Problem Choice: Gender Differences

These general points aside, however, how are the differences in the way men and women choose problems likely to be manifested? In regard to certain fundamental criteria of problem choice, I expected there were going to be *no* differences between men and women in whether these are considered.[11] For example, I expected that there would be equal proportions of men and women giving affirmative answers to the following:

(a) Was the problem chosen suggested by theory in their own field?

(b) Before they chose the problem, did they compare several ideas for research topics and choose the best one?

(c) Before they chose their research problem, did they consider whether the problem has been assessed as important by renowned researchers in the field?

Still, there are other subsidiary criteria on which research problems are appraised where I believe men and women may differ in rankings of importance. Wajkman (1995:202) notes that engineering culture, with its fascination with computers and the most automated techniques, is archetypically masculine. Of all the major professions, engineering contains the smallest proportion of females (see Fox:212-213) and projects a heavily masculine image that Wajkman (*loc. cit.*) sees as hostile to women.

An "engineering" culture may pervade specialties in disciplines outside of engineering itself. Thus, certain specialties involving extensive use of hardware of various sorts may attract men and women differentially. This leads to the following hypothesis:

Hyp. 3.10: **Men more than women researchers will indicate that "access to special equipment that is not always available to me" was a**

criterion for choosing their most recent research problem.

Rossi ([1985] quoted in Giele [1988:312]) noted that "an apparent predisposition in the female to be responsive to people and sounds gives her an edge in communication, whereas males appear to have the advantage in spatial perception, gross motor control, visual acuity ..."[17] These "biological underpinnings of gender behavior" may predict the following:

Hyp. 3.11: **Women more than men will report a desire to work with knowledgeable colleagues was a factor in their problem choice; while**

Hyp. 3.12: **Women more than men will assert that a desire to learn from experts in the field was a factor in choosing to do (collaborative) research and;**

Hyp.3.13: **Men, more than women will claim that a desire to do the work before a rival was a factor in choosing to do a particular piece of research.**

This completes the list of hypotheses that formed the basis for my inquiry.[18]

Summary

The intent of this chapter was to (a) set forth a general strategy for evaluating several models of problem choice, especially in respect to how well they could account for gender differences in problem choice, and (b) state testable hypotheses that would shed light on which model might be the best candidate as a foundation for an empirical theory of problem choice.

The general strategy I proposed called for testing the several models in five spheres where one or more of the models either (1) made specific assumptions, or (2) made predictions or (3) in which implications could be reasonably drawn that could be falsified empirically.

The findings of the various hypotheses stated in the Chapter would bear on the plausibility of the various models and, in many cases, also on why there are gender differences in problem choice. I made clear at the outset that I preferred the "garbage can" model which my literature review in Chapter One had suggested was less vulnerable to criticism on various grounds than other models and which appears to be the best fit with data suggesting there are gender differences in problem choice (a subsidiary concern of this study).

The hypotheses most pertinent to my case that the "garbage can" model is superior as a foundation for an empirical theory of choice are: (1) those that bear on the autonomy of scientists in problem choice, and (2) those that bear on organization culture and gender effects on problem choice.

In explaining the rationale for my hypotheses on scientist autonomy, I challenged the value of any theory of problem choice that depends on an assumption that scientists are free to choose their problems. A theory that makes this assumption cannot be applicable to many settings in which research occurs, *e.g.*, government, commercial and nonprofit organizations which employ research workers. Furthermore, the assumption of researcher autonomy is questionable even in academic settings outside the United States as Shenhav (1985) demonstrated. In contrast, Zeldenrust's "garbage can" model makes no assumption of autonomy and, thus, is superior to the rationalist model in my opinion in this respect.

After pointing out that the Zeldenrust model permits certain underlying regularities in the choice process, I hypothesized that although overall the assumption of autonomy would not be realistic, empirically it seems that it would be possible to imagine fewer women researchers than men researchers enjoying such autonomy based on historical circumstances in academic and other settings.

Since a fundamental assertion of the rationalist model is that problems are almost always theory derived (because theory is viewed as creating areas of ignorance), I argued that evidence of local origination of problems would be significant in demonstrating that a model that predicts problems will be theoretically derived is less empirically correct than one *not* insisting on such theoretical derivation. It is important to understand that simply because it may someday be possible to show all problems can be derived from theory, this is not how problems come to be considered important now, *i.e.*, as requiring solution. A theory that explains problem choice without requiring theoretical derivation of the problem beforehand is, in my opinion, superior to one insisting that virtually all problems scientists work on when free to do so are theory derived.

In regard to competition and conflict, I expressed skepticism that interpersonal competition was a major source of problem choice. I could not, however, rule out the probability that structural competition is relevant to problem choice. And I also argued that competition between organizations in which the scientists work, by its effect on organization culture, could have an *indirect* effect on problem choice of scientists.

Though interpersonal competition in science has received a lot of attention in the literature, I doubt that it is as common as might be supposed, given all the attention it has received. I also doubt that it is a major source of problems. I know of little convincing empirical research results showing widespread interpersonal competition in science. The seductive rationalist hypothesis aside, there are no good theoretical reasons for supposing intense interpersonal competition among scientists.

Structural competition has received less attention, but there are ample grounds to suppose it plays a role in problem choice. If someone offers a large prize to find a cure for breast cancer, that will create a demand that could entice people to do research on this subject when they might otherwise have considered some other projects. (Such structural competition could also cause inter-organizational competition among groups drawn to the prize and, thus, could cause problem choice to be influenced in that manner.).

In regard to organizational culture, I argued that it is a crucial variable in problem choice. (This is, of course, a view not consistent with the rationalist model which sees the characteristics of the problem itself as creating the demand for its solution.) Specifically, by its effect on control strategy, organization culture could steer problem choices in directions more relevant to organization needs for boundary maintenance between the organization and others (or between science and the rest of society) than in the direction of intellectually significant problems.

This departure from "rationality" exemplifies the "blocking action" phenomenon that White (1992) saw as a crucial feature of organized action in his statement of the network theory model.

Finally, I urged attention to gender differences in problem choice as a new avenue to understanding the puzzling phenomenon of gender differences in scientific productivity. I indicated that the differences could take either of two forms: women might be making risky choices hoping to do important work and "striking out" more often than the more cautious male researchers or, women might be choosing problems with great care. By expending much time in their problem search, they might be falling behind their male colleagues who devote less time to problem search and more to problem solving.

ENDNOTES TO CHAPTER THREE

[1]I follow Giele (Smelser, ed [1988:294]) in differentiating gender from sex. She notes that . . . since 1970 there has been a growing consensus on the distinction between sex and gender. Sex more often refers to the biologically determined sex characteristics of male and female; gender, to the culturally and socially defined elements of male and female role expectations."

[2]White sees counter control attempts going on in *every* dyadic relationship and illustrates his point with an example drawn from politics in fifteenth century Florence (1992:88-89).

Commenting on the rise of the Medici faction to supreme power within the Florence of the fifteenth century (based on unpublished data supplied by Padgett and Ansell, Ms., 1989) for example, White observes that while everyone can see that the Medici centralized control, what was less obvious is that they segregated ties to different dependents into different sorts of connections. This kept the dependents relatively separated and segregated, connected only via the Medici themselves and therefore not a good base for success in the *chronic counter control attempts* [emphasis added].

[3]Goode (1973), in a major theoretical essay on professionalization in society, relates professional status to possessing a large corpus of abstract knowledge and a "service ideal." All scientists are members of disciplines that possess the requisite corpus of abstract knowledge. But only university professors among scientists also possess the service ideal. (*see* Goode [1973:354-357]).

[4]Indeed, as new theories take hold, problems that seem to be derived from previous models may, in some cases, no longer make sense in her opinion (see Zuckerman 1978). (Kuhn [1970:102] shows that this is not inevitable; *i.e.,* the problems may be derivable also from the new theory. For example, this occurs in the case of Newtonian model problems which, with certain assumptions, can be derived from the Einsteinian relativity models though, as Kuhn indicates, the terms of Newton's model are understood differently by an Einsteinian physicist than by a Newtonian physicist). In contrast, the constructivist position sees problems as locally derived from the unique interaction of people, equipment and conditions in particular laboratories. These two seemingly contrasting views may possibly both be partly correct.

[5]Aczel (1996:31) reports that scientists have contributed to the war efforts of their society from the earliest days. For example, Archimedes invented engines of war for Syracuse in the third century B.C.E. to help it repel a Roman invasion from the sea. In the seventeenth century, according to Merton ([1973] originally publ. 1935) English scientists conducted numerous experiments directly related to military technology. Merton estimated that (1973:208) ". . . on average about 10 percent of the research carried on by the foremost scientific body in seventeenth century England was devoted to some aspect of military technology."

Scientific personnel participation in these conflicts, however, is not the concern of this essay. It is sufficient for my purposes to note that these data indicate that nonscientist demand for research is often conflict related.

[6]Caplow observes that "Sociological textbooks," beginning with Park and Burgess, "have distinguished conflict from competition and cooperation;" however, he maintains, it is one of the commonplaces of social science that these processes form a continuum," a point he illustrates in his remark that,

> "Some cooperation between antagonists is required in order to stage a fight, and, conversely, a struggle for advantage always takes place when two groups cooperate. Virtually every conflict of organization involves some scarce goal over and above the mutual animosity of the parties. In virtually all competitive situations some degree of hostility develops between the competitors, as soon as they are aware of each other's existence.*"

---

* Caplow's animadversion can, perhaps be illustrated by an example in the history of science described by Merton (1973:335).

> Newton's voluminous manuscripts contain at least twelve versions of a defense of his priority, as against Leibniz, in the invention of the calculus. Toward the end, Newton, then president of the Royal Society appointed a committee to adjudicate the rival claims of Leibniz and himself, packed the committee with his adherents, directed its every activity, anonymously wrote the preface for the published report on the controversy - the draft is in his handwriting - and included in that preface a disarming reference to the legal adage that no one is a proper witness for himself and [that] he would be an iniquitous Judge, and would crush underfoot the laws of all the people, who would admit anyone as a witness in his own cause.

Newton's furious legal maneuverings, notwithstanding, Leibniz shares honors with him as co-inventor of the calculus.

A more recent example in the history of science, however, pinpoints the role played by desires to "scoop" others at work in the field, and so to gain recognition for their accomplishments (Merton:[1973:327]). I am referring to the race to discover the structure of DNA, the chemical building block of life, between Linus Pauling and his assistants at the California Institute of Technology and the Cavendish laboratory in England where Watson and Crick, the co-discoverers of DNA's structure, did their Nobel prize winning research. Merton is properly

For purposes of the present study, only one form of "organized conflict" that Caplow discusses seems pertinent: "continuous conflict or rivalry." Caplow (1964:347) calls this type . . . the most interesting, the most universal and the least understood form of conflict, which usually extends over a considerable period, the timing of aggressive action is relatively unimportant and the means are not specified. The struggle involves realistic goals for each group. Polarization is profound. Strategies are subtle and complex.

Despite Caplow's extensive attention to the issue of "organizational conflict" he did not actually define the term though he criticized earlier definitions.

Kriesberg and Blalock are two writers who did try to define organizational conflict. (Randall Collins (1975) who wrote a book entitled *Conflict Sociology* did not define the term "conflict" at all as far as I could tell.)

Kriesberg (1973) defined "social conflict "as ". . . a relationship between two or more parties who (or whose spokesmen) believe they have incompatible goals (p. 17)." Kriesberg distinguished conflict from competition: Conflict is related to competition; but the two are not identical.

Competition may or may not involve awareness; conflict does. In the case of competition, parties are seeking the same ends whereas conflicting parties may or may not be in agreement about the desirability of particular goals (p. 18). He also differentiated social conflict from other kinds. Situations which an observer assesses to be conflicting but which are not so assessed by partisans do not constitute social conflict. We refer to such situations as objective, potential, or latent conflicts. If the parties come to *believe* that they have incompatible goals, a social conflict has emerged (p. 18).

Blalock (1989) defined social conflict as ". . . the intentional mutual exchange of negative sanctions, or punitive behaviors by two or more parties, which may be individuals, corporate actors, or more loosely knit quasi-group" (p. 7).

Blalock's definition is sufficiently general that it can cover the subtle ways in which a rivalry can be expressed between contending "schools of thought"

---

skeptical that opportunism in science is a peculiarly modern vice. (What else but "opportunism" can one call the efforts of "Simon Mayr (Mariers), who 'had the gall to claim that he had observed the Medicean planets which revolve about Jupiter before I [Galileo] had [and who use] a sly way of attempting to establish his priority?" (Merton [1973:287]). (Bracketed statements are in the original quotation except for [Galileo] which is supplied).

in a discipline and their champions. Problem choices in this kind of conflict can be negative (or positive sanction) used to punish (or reward) other researchers.

[7]However, with costs of equipment and staff higher than in the inexact sciences, I expected that certain specialties within the exact sciences will experience competitive pressures from the budget battles with applicants for funds from *other* fields. This, however, is not as alienating as struggles with one's own colleagues.

[8]I reasoned as follows: Women scientists, by covert and overt means, have been differentially discouraged from pursuing careers in science and within the academic world compared to men (*see* Cole [1979]; Zuckerman [1988]; Zuckerman, Cole, Bruer [1991]; Fox [1995]). Because of past discrimination, women's awareness of competitive behaviors violating the "norms of science" and likely to result in damage to their careers will be greater than men's awareness of these behaviors. (Women may also be victimized, or feel victimized, by these unfair competitive behaviors more than men but I have no way to assess this.)

[9]Zeldenrust offers other interesting hints of the possible importance of organization culture. He notes (1990:307) that in the case of his Netherlands biotechnology research units . . . the investigated research processes exhibited a strong tendency to become *entrenched* in a certain configuration and (thereby) in a certain research theme. Even if in the course of time major changes occur in the environment of the group (*e.g.,* growing industrial interest in the research theme) such niches were not abandoned. A niche was only left under the threat of extinction (or expanded in a situation of great accumulation).

He further observes (1990:292) that in many disciplines ("parts of biology, sociology or the humanities") there are what are known as 'schools.' He avers, schools distinguish themselves particularly from each other through their conceptual development; they are in a sense differentiated, self contained local audiences.

What these comments indicate is the possibility that organization culture not only may help define which among many problems the scientists might select (of those known to them), the organization culture might in some sense shape their understanding of what is a problem because the culture includes sets of cognitions.

[10]Zeldenrust and I focus on processes at the level of the research team rather than the individual. It is useful to do this to demonstrate that problem choice is a result of a process of exchange of resources with the environment. It is not a complete model of the process of problem choice because it looks only at the process from the standpoint of the *team* which implements research after the problem choice is made.

A complete picture of the problem choice process would require analyzing all the actors in the exchange process leading to the decision, *e.g.*, the funding sources, the relevant scientific audiences, *etc.* Theoretically, the conceptual tools for such an analysis exist *e.g.*, White (1992) though it would be costly to implement such a study.

[11]I did not have in my study population a large enough number of researchers from the profit-making sector of the economy to look at this dimension. However, I believe it is relevant to certain aspects of problem choice and I hope I can investigate it in the future.

[12]Gender differences were not originally a focus of attention in this study. Only after I had looked at the qualitative data, especially the material on competition and conflict, did I notice that my women respondents were answering differently than my men respondents. This aroused my curiosity and I began to do some further exploration with simple cross-tabulation of several variables by gender. The results were astonishing. Even more surprisingly, when I checked the professional social science literature on gender differences and on women in science, I found no mention of possible differences in problem choice by gender in any of the discussions of "gender gap" in productivity that I reviewed.

[13]The debate revolves around two related questions:

(a) Is there a productivity gap between male and female scientists?
(b) If there is, what is its cause(s) and implication(s)?

The protagonists are J. Cole (1979, 1991) and his collaborators who find that universities *generally* are merit-based in their promotion policies. They maintain that because of a productivity gap women have not risen into the ranks of senior faculty in proportions similar to their proportions in the overall population. Their position is criticized by feminist scholars on various counts. Ferber (1986),quoted in Frank (1992), has challenged the "validity of Cole's (1979) citation analysis to judge relative scholarly achievements of men and women."

The most recent statement of the position of J. Cole and his collaborates is found in Zuckerman, Cole, and Bruer eds (1991). *The Outer Circle*, New York: W. W. Norton & Co. (reprinted Yale University Press, 1993). Criticism of the latest Cole position focuses on its completeness as an explanation (see Valian[1998]); and absence of empirical evidence (Xie and Shauman[1998]). None of the studies suggests that the problem choice process influences differential productivity of men and women scientists. However, Fox (1995: 220-221 hints at one factor relevant in problem choice, "reported resources," which are differentially available to men and women scientists.

[14]If "relict" behavior in humans is accepted, then the woman scientist need not have a "rational" concern about current acceptability of women scientists in the academic world. It is sufficient that the problem of a frosty reception was real in the past. "Relict" behavior, both in humans and even lower orders of animal life is controversial among biologists and, probably, among behavioral scientists. Yet some research into "relict" behaviors is underway. The *New York Times* (December 24, 1996, p. C1, 6) reported that scientists

> . . . questioned 3- and 4-year olds and adults about childhood nighttime fears. While the overwhelming majority of males reported being fearful of attack from the side, the greatest number of females reported being fearful of attack from below. [One of the scientists involved in the study] says these differences may be due to the life patterns of ancient hominid ancestors. According to some theories, early female hominids were more adept climbers (evidenced today, in part, by the greater flexibility of the young adult female foot) and spent more time in trees than males. More likely to sleep in elevated roosts, females were most vulnerable to attack from below. Males sleeping on the ground, however, would have been more vulnerable to nighttime attack from the side.

[15]There are two ways in which women scientists may, by problem selection, fall behind their male colleagues in productivity. One is to adopt a risky strategy of working in the interstitial areas between their own discipline and others or in other marginal specialties. By this logic, to use a baseball analogy, women scientists are behaving like George Herman ("Babe") Ruth, the legendary "Sultan of Swat" who swung for the fences--and struck out more than most batters. Are women scientists "swinging for the fences" and striking out more than men who are hitting lots of singles and fattening their batting averages?

Or are women trying to be so professional in their behavior that they are (to use another baseball analogy) not swinging at good pitches often enough and thus not making contact with the ball? By this alternative logic, women scientists are overly concerned with their acceptance in the academic world as latecomers to the faculty ranks. They are therefore only doing studies that *everyone* believes are respectable, *i.e.,* spending too much time trying to do their research "the right way." In short, they are not taking the moderate risks on problems that pay good dividends in the form of published papers and professional recognition.

[16]I have deliberately refrained from treating gender as a "causal" variable throughout this discussion. Berk (in Smelser ed. [1988:166-167]) succinctly

addresses the dangers of speaking of non-manipulable variables as causal in the experimental model sense. Nevertheless, he concludes,

> ". . . for many sociological questions, one is not really interested in the impact of gender per se, but some *aspect* of gender that *can* be altered. The same holds for race and age. . . . The point is that the impact of race, age, and gender are *socially* determined."

Berk (1988:167-168) makes an interesting point pertinent to Cole and Singer (1991).

> . . . there are indirect strategies that may be used to obtain some purchase in the impact of race, age, and gender. For example, suppose one believed that women graduate students in the natural sciences are professionally disadvantaged because prospective male mentors are more likely to recruit male graduate students. If one can assign graduate students to senior faculty (within interest areas) on the basis of known covariates, such as performance on standardized tests, by grade point average, or based on the quality of written work, the assignment mechanism is ignorable. Then if the impact of gender on some outcome, such as future publications, is truly operating through differential mentoring, an analysis of the impact of gender, holding constant the covariates, would show no gender effects.

[17]An alternative perspective to Rossi is Leibowitz (1978), at least in respect to *parenting* behavior (not an issue in this essay). Leibowitz argues that "there is great flexibility in parenting capacity and only limited genetic programming of nurturing behavior (her views are summarized in Giele [Smelser ed. 1988:312-313]).

[18]In the “garbage can”-based comparative organizational perspective, the criteria of access to special equipment, desire to learn from experts, *etc.* on which males and females are expected to differ are comparable in role to the critical issues of "technical feasibility" and "organizational feasibility" postulated as bases of problem choice in the L-Z-F version of the rationalist model. In the Zeldenrust model they are components of "resources" that perhaps can be "blackboxed" by the research team.

CHAPTER FOUR: RESEARCH METHODS

"Who is Wise? (not the master, but)
He who learns from all people."
(Ben Zoma Torah Sage)

General Remarks

This chapter discusses the research methods I used in my study of research problem choice. I begin with a statement of the hypotheses I wished to test based on my theoretical perspective as developed in Chapters 1 and 2. This is appropriate because the hypotheses "drive" the research methods used in a study. Any deviation from the ideal method for testing the hypotheses should be for compelling reasons only; the limitations introduced by the deviation should be carefully explained since invariably there is at least some loss of information as a consequence of departing from proper procedures.

The intent in this chapter is to *specify a test* of the rationalist model that can either clearly substantiate the correctness of its assumptions and its conception of problem choice *or* can demonstrate that the alternative view is probably correct that in science problem choice is *non-rational* and *nonrandom*. I maintain they cannot both be correct because the rationalist and the constructivist perspectives are premised on different conceptions of the nature of a problem choice. (I include the neo-constructivist view with the social constructivist view.)

This is less a statement of the similarities of neo-constructivism and social constructivism than a necessary reiteration of the fundamental incompatibility of both constructivist positions with a model [such as the rational perspective] based on the assumptions that (a) scientists have knowledge of the significance of problems prior to initiating research and (b) scientists are free to choose problems. (Interestingly, Shenhav's resource dependence model also directly controverts the rationalist position on this issue, insisting that problem choice is influenced by external forces to the extent scientists depend for resources on these non-scientific audience "Alters." However, unlike the social constructivist model, the resource dependence model makes no assumption whether a chosen problem is derived from theory or locally originated. In this respect, the resource dependence model is closer in spirit to the neo-constructivist position I advocate than to any other perspective).

If I refute empirically both the rationalist assumptions of (a) scientist autonomy in problem choice and (b) a stable problem set, must I accept the constructivist position? The answer to that I believe is affirmative. But I do not have enough evidence from this test to accept a specific constructivist position.

Before I can embrace the social constructivist position, for example, I must *also* demonstrate that problem choice exhibits no underlying regularities--*e.g.*, no relationship to the technical training and theoretical interests of the researcher and no relationship of the chosen problem's rationale to a cognitively validated "core" of knowledge (which the social constructivists do not accept as existing since there is only socially constructed knowledge). *If* I cannot justify the social constructivist view, I am forced to consider the neo-constructivist position as the least objectionable model, since it makes the fewest assumptions while fitting the evidence of empirical research. And it is also consistent with various theoretical traditions in sociology.

Put another way, I can reject the rationalist model and the social constructivist model if (a) *sometimes* researchers behave as if the rationalist model applies-- *e.g.*, they sometimes deduce their problem logically from the prevailing paradigm, or compare problems before choosing one; or choose a problem urged by a respected researcher-- and (b) sometimes researchers behave as if the rationalist model does not apply -- *e.g.*, they derive a problem from findings of some prior research of their own and simply "follow their noses." This case of scientists sometimes acting in the rationalist way and sometimes not, I believe, would be evidence that Zeldenrust's "garbage can" model is the preferred model. This is not to say that the "garbage can" is the best model--only that it is better than its principal rival theories if the rationalist model is correct *sometimes* and other times not. (If I reject both the rationalist and constructivist model, I do not accept the resource dependence model as better because it does not explain why researchers will try to do studies in the absence of demand, hoping for demand to materialize.)

Embracing the neoconstructivist position, however, is not merely a theoretical commitment. If the rationalist perspective is correct in its understanding of problem choice, the sociologist should do panel studies (on panel analysis see, for example, Section III pp. 231-259 of Lazarsfeld and Rosenberg, eds. [1955]). The neoconstructivist model while not rejecting prospective studies would suggest that they must be extremely long term to chart research trails (by extremely long term I mean five years or more). However, the neoconstructivist perspective rejects the observational studies favored by the social constructivists because, while in theory observational methods could work, in the short time span that social constructivists typically work, there is too little opportunity to develop insight into the underlying regularities in problem choice. (Not finding these regularities the social constructivists' "laboratory studies" claim that scientific problem choice is "chaotic".)

Zeldenrust favors archival data supplemented by interviews of informants to obtain the contextual data that he needs to disclose the underlying regularities behind the "chaotic" decisions. In my opinion, however, Zeldenrust exaggerates the dangers of interview data obtained from principal investigators (which he nevertheless uses!) and he also exaggerates the benefits of archival records (the kind he had access to are a unique resource for Netherlands researchers not found in the United States). The latter suppress possibly relevant factors resulting in an embellishment of the true significance of factors on which they do focus. (A good example is the role of organization culture which Zeldenrust does not consider.) Furthermore, these paper records are not reproducible. In contrast to Zeldenrust, I believe that large scale surveys of principal investigators are useful (though possibly even better would be survey studies of 'social circles'[1]). The reason is that the survey is an economical way to get data on (a) the underlying preferences for certain traits in studies, (b) how these fairly stable preferences developed and (c) why they have become part of the research team organization culture.[2]

Yet, survey methods, while appropriate to the present study, cannot answer certain questions validly. There is no way to be confident that questions probing events of the past will be useful in elucidating the role those events may have played in present attitudes. (Much methodological consideration has been given to this issue. However, I am inclined to doubt that events shaping preferences in science are well remembered months or years afterward.)

One possible way to study the development of the culture of the team is a survey of the social circle(s) to which principal investigators belong. For example, Lorain and White (1977:83) suggest that, in principle, cultures of scientific "social circles" can be studied with network methods (a type of survey approach). However, they are primarily interested in describing the entire social circle as opposed to particular dyads--a focus which seems more fruitful in the present example. Since their approach seems to complement rather than replace more traditional sociometric approaches, it is unlikely that any major consequence results from not acting on their suggestion.

While the social circle survey is appealing, one difficulty with studying social circles by survey method is that, unlike formal organizations, they do not have an exact roster of members. Crane (1977:161) emphasizes the "amorphous character" of social circles, pointing out that the scientists who work on similar research problems "... are highly individualistic and widely separated geographically. Participation (in the social circle) is voluntary. Turnover is very high; the majority of scientists have only one or two publications in any research area."

In sum, recognizing that the principal investigator's social circle can be a source of important underlying regularities in her problem choice is one thing. Deciding *which* social circle to which she has belonged is relevant to the problem selection is quite another. Consider a sociologist's desire to do studies that are based on observation methods. Is this principal investigator's desire (exemplified in a scholarly paper) based on observational studies over the past few years, the result of participating in a particular group that includes others who are enthusiasts of that method? Or did they come together to begin with because, among other things, they agree about this method's value in social research? (This is the old "birds of a feather flock together" question.)

The answer is probably that both played a role. That is, (1) a belief that she and they had much in common and could learn from one another played a role and (2) their interactions in discussing finer points of the observational technique of data collection also had some bearing on the problem choice of the sociologist (which seems generally to require an observational method regardless of the topic). However, her original belief in the value of observational research perhaps stems from her participation in a quite different prior social circle (*e.g.* a small group of graduate students in the same sociology department who have since scattered [some may have left sociology altogether] and do not keep in contact with one another).

There are fiercely difficult questions of a procedural nature in assigning attribution in the above example, *i.e.,* figuring out which social circles, *etc.*, contributed which of the underlying regularities in our hypothetical example. Here, while acknowledging that it is possible for me to provide only an incomplete version of "organization culture" (*i.e.*, the respondent's subjective understanding of the organization culture of her milieu), I hope to demonstrate that there is value in this concept to a neoconstructivist hypothesis of problem choice. Before I go further in this direction, I want to return to a discussion of the rationalist model.

Operationalizing the Rationalist Hypothesis

A. Testing the Assumptions

To test the rationalist hypothesis requires a two-stage assessment. In stage one, the sociologist must assess the empirical support for the assumption underlying the rationalist hypothesis: that scientists have control of decisions about the problems to be studied. In stage two, the sociologist must test the empirical support for the hypothesis that scientists, except occasionally, derive their problems from already identified sets of problems based on theory commitments. Furthermore, the sociologist must test the implications of the rationalist hypothesis, *e.g.*, that there will be interpersonal competition between scientists vying to solve the same problems first.

Let us begin with a consideration of the assumption in the rationalist model that researchers have free choice, *i.e.*, control over which problems they will study. How realistic is this assumption?

In Chapter One I showed that strong criticism (that I find persuasive) in the literature was directed at the rationalist perspective's assumption of autonomy (*see* Shenhav [1985]).[3]

My first hypothesis, Hyp. 3.1, is that the proportion of scientists who regard themselves as having great freedom to choose their problems is greater than ninety (ninety-five) percent of the total.

To test Hyp. 3.1, I developed three indicators:

(a) Do scientists report that they had a great deal of freedom to choose problems they wanted to work on?

(b) Do scientists report that the sponsor came to them and told them to do a specific piece of research for them?

(c) Do scientists report that the organization they work

for wants them to work on problems brought to them by their organization superiors?

Each of these three indicators was incorporated in the questionnaire/interview protocol administered to the 107 respondents.[4] (Shenhav used just *one* indicator similar to my question about "great deal of freedom to choose problem." Although I think it is a good indicator, I prefer not to depend on just one).[5]

I can now state the *operational* hypotheses I actually investigated to test the rationalist models key assumption of autonomy.

If the rationalist model is correct, 90% (95%) of scientists:

Hyp. 3.1.1a **Will report that they have a great deal of freedom to choose problems they will work on,**

Hyp. 3.1.1b **Will deny that a sponsor told them what research to do (indicating that the sponsor agreed to fund the researcher's suggested study),**

Hyp. 3.1.1c **Will deny that the organization wants them to work on problems brought to them by organization superiors rather than on problems the scientists choose themselves.**

As stated in Chapter Two, there are grounds to suppose that Zeldenrust is correct in his criticism that the rationalist model presumes a stable set of problems from which to choose, at least over a period of a few years. This criticism led me to state the following hypothesis:

Hyp. 4.1 **Scientists choose their problems from a stable set of already identified problems.**

The rationalist model's possible assumption of a stable set of problems was tested in two ways. First, I asked researchers if they obtained their problems by comparing problems with one another. Second, I asked the respondents if they knew what their next problem would be.[6]

In regards to whether researchers compared problems to one another before choosing, I asked this not only about problems in particular, as in Q22 (see Appendix "Questionnaire") but also about "topic." I asked further whether they had compared problems in the past before choosing one.

My focus on problem and topic comparison was the result of my conviction that only when there is a *stable* set of identified problems can comparisons occur to establish which problem is more significant and/or more feasible to study. An absence of prior comparison indicates either that no such stable set of problems exists or, less likely, that the researcher is rather idiosyncratic in her problem selection. (Clearly, when information is available, one uses it to reduce ambiguity unless one is using an idiosyncratic selection procedure).

The actual hypothesis investigated is:

If the rationalist model is correct, scientists:

Hyp. 4.1.1a **Will report that they obtain the problems they will research by comparing problems with one another,**

Hyp. 4.1.1b **Will report that they knew what their next problem would be well before they started to work on it.**

The results of the empirical test of this hypothesis of the rationalist model are found in Chapter Five (*see* Table 5.2). They confirm that there are a substantial proportion of scientific research workers (49%) in senior research positions who do not compare problems first before choosing a problem majority. They also show that most do not know what their next research problem will be. In respect to the latter finding, it seems absurd to believe that my respondents did not want me to think they knew beforehand what their next problem might be (assuming they already knew what they would study).

B. Testing the Hypothesis Itself

Now that I have addressed the assumptions of (1) autonomy in problem choice and (2) stability of the identified problem set, I can turn to matters more directly bearing on the rationalist perspective.

I believe that it is not sufficient to falsify a hypothesis by questioning assumptions underlying it. It is possible that other unknown assumptions are correct and so the hypothesis may still be correct. Thus, I needed to test the rationalist hypothesis itself and its implications in addition to challenging its assumptions. The hypothesis itself asserts that problems are chosen from among the identified problems except occasionally, and that this choice is based on theory commitments and other assumptions. (In the context of the article by Zuckerman, these other assumptions are certainly methodological in character and not related to practical, political or other considerations.)

Testing the rationalist hypothesis is not so simple as it sounds. Just what is meant, for example, by the qualifier "except occasionally" in Zuckerman's hypothesis? Obviously, Zuckerman meant that the vast majority of serious researchers derive their problems from available paradigms. Yet, she allowed that once in a while, scientists would do otherwise and this would still be "rational."

I decided that operationally, Zuckerman's hypothesis required that scientists would have to choose problems in the way she expected them to either *ninety* or *ninety-five* percent of the time.

I settled on these criteria of the ninety and ninety-five percent levels because, depending on how strictly one construes "except occasionally," either seems to be consistent with Zuckerman's intent. It is this strong assertion of the fundamental hypothesis of the rationalist model that makes it a good candidate for testing. It can be *falsified* unlike so many sociological hypotheses that vaguely refer to a propensity of one sort or another.

However, now I also needed to look at indicators of choice based on "theoretical commitments." Shenhav (1985) implicitly argued that it would be sufficient to falsify the rationalist model by showing that problems came from "external client suggestion" primarily and not from "the literature." However, I have several objections to his suggestion. To begin with, I contend that his indicator lacks apparent validity[7] because choosing problems from those already identified is *not* the same as choosing from the literature. Zeldenrust (1990:7) illustrates the limitations of "the literature" as an indicator when he quotes Crane, referring to the rationalist perspective, that "problems are seen as derived from existing nested bodies of accepted knowledge, which are guarded by elites or invisible colleges (Crane, [1972])." The crucial point here is "accepted knowledge," and "the literature" includes both accepted and not accepted ideas.

Furthermore "the literature" is too vague a concept in an era where much information circulates in the form of electronic mail. Even assuming, however, that "the literature" could be operationally defined, it is possible that "the literature" as a sole indicator excludes other informal means by which problems of significance are disseminated, stored and retrieved. Especially in the case of recently found problems, the only material available may be notes from conversations among researchers. Counting such notes, e-mail messages passed from one member of the 'invisible college' to another, or even draft articles circulated among a limited group of senior scientists and their associates as "the literature" is to make the term so elastic as to render it almost meaningless.

Consequently, Shenhav (1985) cannot be said to have disconfirmed the rationalist perspective with his finding that researchers are often more influenced by clients than by "the literature."

Three other indicators specified below, I maintain, are central to the veracity of the rationalist hypothesis because they bear on how scientists, who are free to choose, will derive their problems. Those indicators are:

1. Problems chosen are suggested by theory;
2. Problems are chosen after work on them is urged by respected and/or renowned colleagues in the scientist's discipline and
3. Problems are usually chosen after comparison.

I shall address each of these three indicators in turn. (The reader will note that the indicator of whether choice occurs after comparison also bears on the *assumption* of a stable problem set I believe may be immanent in the rationalist hypothesis. Therefore, by challenging the evidence that comparisons of problems are made before problems are chosen, I would be attacking both the hypothesis of rational choice *and* the assumption of a stable set of problems from which to choose that I believe is implicit in the rationalist model).

1. Problems chosen are suggested by theory

Zuckerman (1978) clearly saw problem choice as indicative of certain "theoretical commitments." She notes (1978:82) that scientists confronted by "arrays of previously identified problems" used "assessed scientific importance of a problem" as one of the two most frequently employed criteria in selecting problems and this desideratum "of course reflects *theoretical* commitments" (emphasis added). Undoubtedly, Zuckerman had in mind that the *significant* problems in the stock of identified problems were those suggested by *theory* in the field. Other problems might be found in “the literature" but mere presence in the literature was not the relevant issue. The problems should be deduced from the controlling theory (*i.e.,* one of the main paradigms) in the discipline.

2. Problems are chosen after renowned researchers and/or respected colleagues urge work on them.

In some disciplines such as medical research it is unusual to find powerful paradigms from which researchers can derive problems. Respected colleagues' opinions of what is important may play an important role in problem choice in fields without paradigms.

I regard research undertaken after checking with respected colleagues to be problem choice in the rationalist fashion. Therefore, I believe this indicator is a necessary supplement to the first indicator which points to "theoretical commitments."

Consequently, I state the following operational hypothesis:

Hyp. 4.2: If the rationalist model is correct, scientists will report that the latest problem they chose was suggested by a theory in my own field or after respected colleagues urged work on them.

This hypothesis, if supported by the data, would suggest that problems investigated can be related to the paradigm of their discipline. However, even if the hypothesis is true, it does *not* confirm the rationalist model. But it does throw into doubt the social constructivist claim that (a) problems are always locally originated and (b) that they are socially constructed. This is because it underlines the fact, that (a) there is a *cognitive* content to the problems, (*see*, Giere [1988]) and (b) hypotheses are usually related to the "core" (*see*, S. Cole [1992]) of the discipline. And this core is not merely socially constructed but is "real" in a meaningful way.

Therefore, the merit of this hypothesis, if correct, is to enhance the credibility of a neoconstructivist position, such as Zeldenrust's, relative to the social constructivist position. When *taken together with the findings on autonomy and the stable problem set from which choices are made*, moreover, it also casts doubt on the rationalist position regarding the process of problem choice.

3. Problems are usually chosen after comparison

It is intrinsic in the idea of a "stock of identified problems" (see Zuckerman [1978]) that scientists will *compare* problems. Zuckerman (1978) made this clear when she spoke of the use of "assessed scientific importance" as one of two most frequently cited criteria for evaluating problems presented in an array.

In a modified version of the rationalist perspective with which I began this inquiry, (the Lederberg-Zuckerman-Fisher model), the comparison is more involved perhaps than suggested in Zuckerman (1978). The scientist is seen as weighing various problems against three broad sets of criteria: (1) technical, (2) organizational, and (3) political/legal.

The need, in my opinion, to presume such involved decision making arises from the advent of "Big Science." Scientists now conduct studies that cost millions of dollars; that require large teams of researchers and expensive equipment; and that take years, even a decade or more to be completed. The idea that these studies are considered with great care prior to their inception is not implausible. Yet, the emergence of "Big Science" has not led to the demise of "Little Science.". Big Science and Little Science exist side by side. There are plenty of scientists who do modest studies costing perhaps $30,000-$50,000 that make important intellectual contributions to their respective disciplines. And the idea that a detailed comparison of numerous problems in three broad *classes* of criteria occurs before a modest budget study commences seems farfetched. Either the comparison process is much simpler than supposed or it is absent in "Little Science."

The "garbage can" model of Zeldenrust suggests that the decision process is indeed simpler in "Little Science." His model borrows heavily from the constructivist "laboratory" studies the idea that choices are not made from a list of identified problems. Though there are underlying regularities, the actual choices are the unpredictable results of linkages made among three (sometimes four) flows: (a) demand, (b) problems, (c) resources and sometimes (d) constraints. The Zeldenrust model makes no assumptions about the provenience of the problems; they can be derived from theory, they can be locally originated or whatever.

In Chapter One, I already indicated that the rationalist model failed to account for the large percentage of scientists in my respondent group who originated problems locally. It is important to understand that *perforce* those scientists locally originating their problems were not comparing problems drawn from theory, paradigms, *etc.* prior to doing research. The significance of this finding, from the methodological standpoint of the present chapter, cannot be overstated. If the rationalist model were correct that prior comparison of problems was done almost always before a choice were made, it would also indicate that problem choice involves selection from a *stable* stock of already identified problems - an assumption crucial to establishing the superiority of the rationalist model.

And, if the assumption is valid, a particular research procedure for study of choice is required. Essentially, given the validity of the assumption of stability, one should study choice much as Berelson, Lazarsfeld, McPhee (1966) and Lazarsfeld and Rosenberg (1955; Part III) studied voters' preference shifts before they actually cast their ballots. Berelson *et al.* method was one of *repeated* panels of interviews and study of the large social processes that impinged on the voters over time.

However, if the assumption of a stable problem set is not valid, then repeated panel analysis will only demonstrate the unpredictability of the choices. Knowing a preference for doing one study at one point will reveal nothing about the decision regarding what to study at the next point. However, a sense of the *underlying* regularity behind the seeming "randomness" can be gotten by interviewing the scientists regarding *characteristics* of desirable problems. These kinds of preferences are stable over time even though which particular problem scientists prefer will not be.

My finding that a large percentage of my study population did not choose their problems after making comparisons among problems for feasibility or importance suggests that cross-disciplinary research is needed to study problem choice. (If my study had shown that problem choice occurs after such comparisons among problems are made, a study of just one discipline's practitioners would have been adequate to understand the choice process.)

The need for cross-disciplinary research adds to the methodological complexity of the research. A truly adequate cross-disciplinary study would also demonstrate that problem choice of scientists is not like the behavior of voters deciding amongst a small stable set of candidates. However, since different disciplines have different rates at which studies of one set of problems are completed and new problems tackled, *as a practical matter*, a cross-disciplinary study that is longitudinal was simply not feasible for this project.

In planning my research based on the presumptive correctness of the rationalist model, I expected that I could focus on chemists who I knew do three or four new projects within a year (*see* Hargens [1975:61]; also *see* Gieryn [1978]). Therefore, I could hope to observe problem choice behavior among chemists longitudinally in a relatively short period. But since problem choice is not done rationally and since studies are completed at different rates in different fields, this longitudinal study is not practicable in a study across disciplines.

A. Testing Implications of the Rationalist Hypothesis

Another area where I needed to develop suitable indicators was the issue of competition and conflict. Competition and conflict are sensitive topics in science. Mulkay (1991:10-11, originally 1974) reports that when he emphasized rivalry rather than cooperation in a study of radio astronomy research groups in England, he received "strongly negative comment, especially from group leaders."

Scientists seem reluctant to discuss their interpersonal conflicts with fellow scientists when talking to outsiders. For example, Zuckerman (1977) found that her Nobel laureates would evade questions about such matters. Oswald Hall (1944) had noticed that his physician respondents also hesitated to discuss such issues with him. Under the circumstances, I thought it best not to inquire into this issue directly.

Instead, to address the issue of conflict, I decided to try indirect indicators. I have two such indirect indicators for interpersonal conflict.

Q: **In your opinion do some researchers choose problems for research *primarily* to embarrass their rivals or personal enemies?**

Q: **In your opinion, do some researchers select their problems because they want to support the intellectual positions of their friends?**

For competition, I tried three indicators (two direct and one indirect measure). These were:

Q: **In your most recent project, was a desire to do it before a rival did it one of your reasons for choosing the research problem?**

Q: **Has the desire to do this before someone else does it been a factor in your choosing research problems in the past?**

Q: **Do other researchers choose their problems based on the desire to 'do it before someone else does it'?**

I reasoned that competition is not something that researchers are ashamed to admit since at least in the popular imagination one can have a "friendly" rivalry (though Caplow [1964] insists that some kind of hostility is present in any competitive relationship in which one is personally acquainted with one's competition).

In Chapter Two I presented my operational definition of the concept of organization culture as a three dimension variable. However, because of the small size of the sample, especially the commercial researcher component, I was not able to implement a study of organization "climates." Instead I focused on the impact of specific cultural variables.

With reference to gender, I did not originally intend a study on this issue. Consequently, I had to depend simply on a self report question regarding the gender of the respondent.

B. Data Collection Procedures

Now that I have discussed hypotheses and operational definitions of certain terms important to my thesis, I turn to the procedures I used to collect data about the variables of interest. I first describe what procedures I used to choose my case and the characteristics of the cases resulting. In a Technical Appendix, I lay out a justification for my unconventional procedure (which is not unusual for researchers studying special populations).

This study is based on data from a nonprobability ("convenience") sample of 107 scientists, who either responded to a lengthy questionnaire or were interviewed (by telephone or in person) using the same questionnaire.[8] As the tables below indicate (see section on Characteristics of Study Population), the respondent population is a diverse group--many different disciplines, work environments (*e.g.*, academic, government, commercial), age ranges, both genders. While not shown, two different countries--Canada and the United States--are represented with Canadians (strictly speaking Canada-based researchers) accounting for about 20-25% of the study population. In the Appendix to this chapter I discuss a statistical/methodological rationale for using such a strategy.

The study is not the first to be done on this topic and a bit of commentary on its distinctiveness is perhaps in order at this point. The chief distinction of my research is the effort made to gather data from respondents in many different disciplines. Velho (1990) and Busch, Lacy and Sachs (1983) focused on agricultural research specialists (though the latter actually did study many different disciplines, all concerned with agriculture in some way). A second distinction is the organization perspective of the researcher. This I share with Zeldenrust, though his study (a) was based primarily on archival data for a limited number of research units and (b) was oriented toward examining different aspects of the problem choice issue such as "ambiguity" and "control strategies."

My study, which brings to the study of problem choice a comparative perspective and a focus on such issues as competition and organization culture (and also gender) supplements Zeldenrust's study.

Though I largely share Zeldenrust's perspective, there are some differences in our points of view. Zeldenrust differentiates "demand" from resources. I follow the economists in that I make no distinction. Demand is *real* to the extent there are resources provided, *i.e.,* demand is coupled to resources. Since the researcher herself is part of the scientific audience, I count the demand from a scientific audience as real to the extent the scientist (or her scientist colleagues) marshalls resources for the study of issues that are scientifically interesting. (It does not matter to me that the resources marshalled were legally intended for some other purpose, a practice referred to (by Laporte and Wood [1970]) as "bootlegging".)

A second distinction is that I do not use Zeldenrust's concept of "driving force" preferring instead to look at "organization culture" which I contend is a propulsive force moving the researchers toward (or away from) work on certain problems (and accounting for differences among teams in the driving force).

There is also a methodological disagreement between Zeldenrust and me since Zeldenrust claims to foreswear research on individuals (though he makes use of them as "informants") whereas I am quite comfortable with research based on individual respondents' answers to my survey.

C. Characteristics of the Study Population

I strove for heterogeneity in order to understand how differences in certain background variates might influence problem choice strategies. Consequently, I sought respondents of both genders, varying levels of experience - though I wanted experienced people, many disciplines as evidenced by highest degree attained. I was able also to achieve some comparability with a large National Science Foundation study population (*see* Table 4.1).

Table 4.1 shows the distribution on two background variates of scientists with earned doctorates as presented in a National Science Foundation (NSF) study performed in 1993 compared with respondents in my study population.

The NSF study is a labor force survey designed to obtain information on the characteristics of American scientific and engineering personnel. Thus, its purpose is not at all similar to that of my study. The results disclose, however, a striking degree of similarity on the two dimensions that are common to both surveys: gender and discipline.

Table 4.1: Comparison of Study Population With Population in National Science Foundation Survey of Scientists and Engineers

Disciplinary Background[A]	*Study of Problem Choice in Science*	*National Science Foundation Study*
Computer/Math/ Engineering	24% (26)	27% (98,620)
Biology/Life	33% (35)	26% (92,180)
Physical and Other Natural Sciences	20% (21)	20% (72,830)
Social/Behavioral/and Related	23% (25)	27% (96,030)
Gender[B]		
Men	61% (65)	70% (288,910)
Women	39% (42)	(30%) (70,760)

[A] National Science Foundation, *Characteristics of Doctoral Scientists and Engineers in the United States: 1993* Table 3, pp. 9-10.

[B] National Science Foundation, Women, Minorities, and Persons With Disabilities in Science and Engineering: 1996 Table 1-1, p. 3.

Table 4.2: Description of Study Population

Work Environment	*Frequency*
Public university/college	23
Private university/college	17
Government	39
Private not for profit	6
Other private for profit	8
Subtotal	93
Missing/System Missing	14
Total	107
Field Working In	*Frequency*

Biological sciences/medicine	35
Other natural sciences	21
Engineer/applied math/math	26
Social, behavioral, sciences, economics	25
Total	107
Gender	*Frequency*
Male	65
Female	42
Total	107
Years of Research Experience Parameter Values	
Number valid	98
Minimum	1.0
Maximum	50.0
Mean	23.81
Median	24.00

4. Validity

The present study confronts some challenges in dealing with the question of validity. I had indicated earlier that there was no practicable way to do a multi-disciplinary *longitudinal* study -- the most valid way to assess problem choice by scientists in a single study. However, the concurrent validity of many of the findings of this study can be assessed in light of findings of other studies done differently or with different populations. The study's findings regarding limits on the autonomy of scientists with reference to problem choice are consistent with findings of historical studies, *e.g.*, Westfall (1977); archival research, Zeldenrust (1990); and sample survey research, Shenhav (1985)and Nederhof and Rip (n.d.).

Likewise, the findings on local origination of problems, especially in medicine and biological sciences, are consistent with Knorr-Cetina's findings (1981) using anthropological field work methods; Zeldenrust (1990), using archival data; and perhaps with Woolgar and

Latour (1979).

The validity of findings on gender-based differences, reported in Chapter Five, cannot be ascertained without further research. No prior research of any sort exists with which to compare these findings; indeed, no one has so much as speculated on the possibility of such a difference between the genders (though my findings are adumbrated by earlier work by Feldt (1986) reported in Fox (1995). I believe these differences, which were serendipitous in my own study are of great theoretical importance in light of the findings on gender based scientific productivity differences. Moreover, if better data are available to corroborate them in the future, these gender-based differences lend credence to the "garbage can" model of problem choice, while undermining current rival hypotheses such as the rationalist model.

Validity of data on competition can be assessed against other studies. Note that Zeldenrust (1990) found no interpersonal competition. Mulkay (1991, originally 1974) believes he did but his evidence is certainly not persuasive on that point. He concedes, furthermore, that the policies of the funding source (U. K. government) prevents *structural* competition which, if it had existed, might have spawned interpersonal competition and/or interpersonal conflict.

I know of no comparable studies on cultural variables (organizational culture or gender) in addition to my own study. Work group studies have a long history beginning with the "Hawthorn" studies of the 1930s reported in Roethisberger, F. G. and Dickson, W. J. [1939] *Management and the Worker*, Cambridge, MA: Harvard Univ. Press. However, there is no data that I know of pertaining to scientific problem choice by work groups.

TECHNICAL APPENDIX TO CHAPTER FOUR

This appendix considers various issues of a general procedural nature that are problematic because this study looks at perceptions and behavior of an elite population. The methodology of survey research on such populations is poorly developed and, I maintain, the general survey method based on random samples is not adequately robust to be useful for this special population. I first consider the special constraints that apply to elite research and then discuss how I addressed them in my qualitative study.

Sample Strategy

Until this point, I have focused on the hypotheses, the indicators and a rationale for them. I also described the sampling method and population. However, because this study does not employ a conventional research procedure for data collection, a rationale for decisions to depart from the conventional design is necessary.

Probability Sampling Versus Non-probability Sampling

Many sociologists believe that the arguments in favor of a probability sample are so powerful that when they see a study based on a non-probability sample, they regard it as a tocsin warning them that an inferior method of research has been employed and scientifically dubious results have been obtained. However, the superiority of one method of sampling over another is not established beforehand; the correct procedure to use in a particular instance is determined by the facts in that particular case. As Kempthorne (1961:124) says, "Scientific life is no different from life in general in requiring compromises which turn out to be good."

The case for a probability sampling procedure is based on a pure statistical concern with rigor and precision. Chen (1990:222) observes that as developed in the work of Sir Ronald Fisher, the classical experimental approach to research founded on the use of probability sampling "is rigorous in mathematical reasoning and fits well with existing statistical theories."

It is not surprising, therefore, that probability sampling methods are extensively employed in large scale social survey research where precision of the parameter estimate is especially important. Typically, large scale social surveys are performed on important policy related issues or in market research where the trustworthiness of the confidence intervals and/or parameter estimates is crucial to the survey users (*e.g.*, government or commercial organizations). I maintain, however, that despite the deeply entrenched support within the social science profession for the probability sample, the intellectual basis for this allegiance is not nearly so strong as might be supposed.

First, I would like to point out that in some sociological specialties, *e.g.*, historical sociology and ethnomethodological research, the sample survey is almost never done. Indeed, the cornerstone of the sociology of science, Merton's various investigations (1935,1973), was not based on research using probability sampling methods.

Second, I maintain that even within the ranks of survey researchers there is no inflexible commitment to the use of probability sampling regardless of the nature of the research. Kalton (1983:91), for example, asserts that probability samples are less important in small sample studies than in large sample studies. His argument, firmly grounded in the classical tradition, is that in the small sample study bias is less likely to be important in relative terms in causing the imprecision of the sample estimate than the smallness of the N of cases itself.

In general, I contend that a rigorous approach to deciding what kind of sampling strategy to follow would require answers to three questions:

1. Is the precision of the estimate crucial because of the costs of an imprecise estimate in human lives, in health, or in significant monetary terms?
2. Is there prior information on the parameter values based on research using different populations and different techniques from those in the current study?
3. Is it feasible to draw a probability sample and collect data from it that are generalizable to a meaningful population?

I address each of these questions in turn.

1. Is the precision of the estimate crucial because of the cost of an imprecise estimate in human lives *etc.*?

The general purpose of a probability sample is to obtain precise estimates of the parameter values assuming that there is no prior information about the variables of interest. From the Bayesian theoretical standpoint, to which I adhere, when a probability sample has been drawn only the degree of uncertainty is reduced by the findings. Iversen (1984:11) says ". . . in order to learn about the parameter we collect sample data. As long as we have only sample data, we cannot hope to find the exact values of the parameters but we know more after the sample data have been studied than we did before."

Bayesians criticize the use of classical statistics to attempt to confer on a study the status of having definitively found the parameter value of some variable of interest. As Iversen (*loc. cit.*) says, "[Classical] theory is based on probabilities as long run frequencies. The significance level [of statistical tests such as the Student's "t" test] tells us what will happen in the long run, if we draw a large number of samples."

Thus, by Bayesian logic, the single probability sample study's external validity, especially that of a small study, is not far superior to a nonprobability sample's external validity. Both the probability sample study's results and the nonprobability sample study's results can be used to reduce uncertainty about the true value of the parameter but neither is definitive in its parameter estimates.

Finally, in respect to the first question, the imprecision of the single nonprobability sample study may not be important in the present case because there are no costs in human lives, health or significant monetary terms associated with an incorrect finding.

Before I address the second question, I would like to make another point about the wisdom of doing a large sample study. Absent a theoretical model that is a plausible alternative to the rationalist and social constructivist models, it is unlikely that this statistical study would change any minds among sociologists. Rationalists object to a view of scientific research as chaotic and unrelated to the core ideas of the discipline.

They would not embrace the social constructivist model simply because it were shown that less than ninety percent of problems were theory derived. Social constructivists for their part, would not embrace the rationalist model simply because a statistical study showed that problems are usually theory derived. Thomas Kuhn (1970:77) makes abundantly clear why this is true. "Once it has achieved the status of a paradigm," he says, "a scientific theory is declared invalid only if an alternative candidate is available to take its place." He continues, "No process yet disclosed by the historical study of scientific development at all resembles the methodological stereotype of falsification by direct comparison with nature.... The decision to reject one paradigm is always simultaneously the decision to accept another and the judgment leading to that decision involves the comparison of both paradigms with nature and with each other."

My study advocating a new model is more persuasive than a simple test of the rationalist or social constructivist models would have been even if that test were based on a large probability sample. I show the rationalists that there is an alternative model to the rationalist model that (a) argues that the choice process has regularities, (b) does not assume that there is no core knowledge, (c) does not dispute that problems can be derived from this core knowledge in most cases and (d) that some forms of competition can be relevant to problem choice.

I also show the social constructivists that there is a view other than theirs that makes clear that (a) problem choice is not a problem driven process nor even a rational process, (b) problem choice is affected by demand from nonscientists and (c) problems may be chosen without regard to any core theory and in fact the choice may be based on entirely local factors.

2. Is there prior information on the parameter based on research using different populations and different techniques

from those of the current study?

The findings of my non-probability sample study done by questionnaire and interview are being compared to findings of other studies done using very different methods, *e.g.*, archival methods and observation methods and with totally different populations. Because of the different threats to validity associated with different methods, if similar results are obtained using different methods and different populations, some confidence that the estimated parameter values are approximately correct is justified. (Note that Bayesian logic *requires* prior information about a parameter being estimated, but it does not require that this prior information be obtained in any particular way.

However, in evaluating whether a non-probability sample study is justified, it is a germane consideration that prior studies have been done on different populations and using different methods and that their results are consistent with those of the present study.)

My research focused on five areas: (1) autonomy of scientists, (2) local origination of problems, (3) interpersonal competition as a source of problem choice, (4) organization culture as a source of problem choice and (5) gender differences as a source of problem choice. For each of the five areas, except gender, I have prior research with which to compare. Thus, in the case of interpersonal competition I showed in the literature review that there was reason to doubt that much interpersonal competition exists between scientists (see Hagstrom [1965]; Zeldenrust [1990]; Mulkay [1991]). Consequently, it was unlikely in my view to be a source of problem choice. However, the rationalist model *strongly* implied that interpersonal competition between scientists should be considerable and might, therefore, motivate which problems of those shared would be chosen to research first.

Similarly, I suggested that in their major exploratory study, Busch, Lacy, and Sachs (1983) had pointed to a possible organizational cultural variable as explaining problem choice (Factor V) in their factor model. Gender, the fifth area I investigate, has not previously been suggested as a source of problem choice differences. However, I indicated in Chapter Two that one could have derived hypotheses for a role for gender from prior research (*e.g.*, Feldt [1986]) and theory. Thus, though a role for gender was an unexpected finding in my study, it is not necessarily an illogical possibility in my model (though it is absurd in the rationalist model given the assumption of perfect information about significant problems).

3. Are data being collected from subjects or about a topic which

cannot feasibly be studied by random sampling designs?

Chein (1976:534-540) offers a number of grounds that may justify doing a non-probability study. First, he notes that "Sometimes there is no alternative to non-probability sampling." He cites the need for data on a closed society such as the People's Republic of China at that time. As he puts it, "The choice here is between data that do not permit a statistical assessment of the likelihood of error and no data at all" (1976:536).

Another rationale would be a situation of choosing between skimpy data from a random sample and rich data from a biased sample of knowledgeable and articulate informants. He gives the example of studying juvenile gang members directly by probability sample versus interviewing a select group of social workers who have insights from their work with gangs.

A third justification which he considers "special and controversial" is data collected from accidental samples of subjects to which statistical tests of significance are supplied.

Chen observes that,

> If we have no special reason for wanting to estimate the degree and character of relationships in an already specified population, it may be easier to begin the quests on non-probability samples and to make use of the fiction of hypothetical populations of which our samples are quasi-probability samples to provide guidelines (for example, statistical test of significance) for the evaluation of findings.

The second rationale seems most appropriate to the present study. I had a choice to make between nonrandom sampling and randomly sampling researchers in all fields. If I chose the latter, undoubtedly, given a small "N of cases," my sample would be virtually all male (since men predominate in science and engineering), and would contain virtually no especially distinguished people (since they are by definition only a small fraction of the research population). Furthermore, sample attrition (and the need to constantly draw by replacement) would have hindered the validity of the study or stretched out to an unreasonable extent the time it took to get a sufficient number of cases. (Scientists are extremely busy people and even my non-probability study -- based on people to whom I was referred as especially knowledgeable and willing informants -- required a lengthy data gathering period.)

Finally, I probably would have missed out on the opportunity to include Canadian based scholars, and senior women scientists in my study population if I had tried to do this study by random methods. This is because constructing a population of elements from which to draw randomly such informants would have been fiercely difficult and certainly immensely time consuming.

And, what would be the point of such an exercise? As Bayesians point out, a single study's external validity is limited. Furthermore, as classical statisticians concede, in a small study bias is not the chief threat to external validity. In sum, given the difficulties of population construction, given the prior information from other studies with which to compare my findings, and given the low costs of an imprecise indicator, a non-probability sample study made eminent sense.

While a non-probability sample seemed to make more sense than a probability sample, it did not solve all of my problems. I wanted to get a wide variety of scholars to allow me to see if findings were affected by disciplinary background (my results show that disciplinary background is an important factor in whether problems are likely to be selected from received theory or locally originated). I also wanted men and women in my sample to be certain that my findings were true of *scientists* regardless of gender (I did not suspect until my data were collected that gender would prove to be an important predictor).

Another criterion that was important initially was my desire to have many people with multi-disciplinary research backgrounds. Though I started out believing these kinds of people might be rather rare, I found that actually many scientists had such experience.

Further, I wanted answers to questions about interpersonal conflict and competition that I believed would be regarded as sensitive. (Actually these questions did not seem to cause embarrassment or evasiveness. The researchers in some cases, however, were sensitive about questions regarding politics in the grants process that I also asked my respondents. Results of those inquiries are not included here.)

Aside from these questions of how to choose a probability sample, I also had to address the issue of developing a strategy to assure access to my respondent population. Scientists and researchers are extremely busy professionals; moreover, all of those I sought to include are higher social status than I. Either they are (a) tenured faculty, often at prestigious research universities or (b) senior scientists of major government laboratories or corporation research units. In some cases they also are highly prominent in their fields; for example, a few are members of the National Academy of Sciences and many hold endowed chairs in their universities.

The prestige and superior social status of many of these scientists is of more than passing interest because it may have had consequences for my respondents' willingness to discuss certain sensitive issues among which are interpersonal conflict and competition in science, subjects which I believed at that time to be important in problem choice.

The methodological literature shows that the issues of individual conflict, especially, and of "politics" in the grants process are, for many scientists, sensitive issues and there is some resistance to discussing these issues with an outsider, especially one of lower social status.

As Oswald Hall (1944) noted in his study of the medical profession, several peculiar difficulties are encountered in trying to study the medical profession. The status of its members is generally higher than that of the persons making the studies. Since it is usually considered inappropriate to discuss one's important affairs with those of a lower status this limits the kinds of facts revealed.[9]

Zuckerman (1979:273-4) reports that in her study of Nobel laureates, some questions were perceived as threatening especially those that seemed to require the laureates to disclose detail of conflicts between co-workers or that reminded them of distressing events in their own careers. It was possible in some cases to reassure the laureate that he was not being asked to reveal confidential information about particular people or events.

Zuckerman also admits there were other occasions, however, when the laureates made it clear that they would not answer a particular question:

> **"Int: You said that Dr. X was not the easiest man in the world to work with. What exactly do you mean?**
>
> **L: I don't know that I'd like to talk about that."**

Questions on 'relative' contributions of scientists to particular discoveries were sometimes sensitive," Zuckerman adds.

In conclusion, Zuckerman found that "On the whole, the laureates have learned to evade, without much embarrassment, questions that they do not wish to answer" (1977:274).

Mulkay (1991; originally 1974) found that in the United Kingdom competition at the national level (but not the international level) was a sensitive issue for his study population of British radio astronomers.

Mulkay and his colleague (1991:10), after interviewing researchers including "'group leaders'," says [that they]" . . . decided to check our work as far as possible by sending the first draft of our report for comment to all those we had interviewed . . ." [because] "of the complexity of our subject matter and the imponderables involved in even the simplest kind of interpretation." He received back "strongly negative comment, especially from group leaders . . . about our discussion of competition/cooperation, leadership, and secrecy."

Mulkay believes that the force with which the research group leaders reacted to his analysis reflects a perception that Mulkay's research was putting in jeopardy the image of science as a cooperative and dispassionate pursuit of knowledge. From this perspective, an account which threatened this image would be seen, particularly by scientists responsible for maintaining the groups' financial and social support, as endangering that support.

The fears of Mulkay's respondents may be quite correct in the British case. The British government had made efforts, according to J. Gaston (1973), to prevent overlap in research and competition; even the hint of it could be interpreted in government circles as bad faith on the part of the scientists dependent on British government support. Thus, Mulkay may indeed have been causing serious problems for his respondents by his data interpretation. Should Mulkay have elided the material on competition? This is an ethical dilemma for the sociologist and I do not know if there is a correct answer. It seems that Mulkay chose to stick to his guns. However, had I done that study I would have given little credence to the evidence of competition because emphasizing the idea of competition could have serious negative consequences for his respondents given that Mulkay's definition of competition was imprecise and subject to misinterpretation.

Another problem in studying an elite population bearing on access to respondents also was discussed by Zuckerman (1979:274, originally 1977). She noticed that the phrasing of questions also seemed important. The laureates responded with unusual precision to the wording of questions. For this reason, it is difficult to use the routine procedure of beginning with a general question and moving successively to more concrete issues. The laureates frequently asked for definitions of terms or supplied complex variation on the meaning of particular wording.

It was obvious to me from Zuckerman's observations that I would jeopardize my entry to the study population I had in mind if I did not prepare a carefully crafted questionnaire. Pretests of the sort of survey researchers usually do, however, were not really an option because it would be difficult enough to find appropriate subjects for the study, let alone for both a pretest and the study.

Once I had developed a questionnaire (and had it critiqued by friendly scientists with whom I and/or my brother Dr. Joel L. Fisher are acquainted), I was ready to try it out on my respondent population. I now faced the hurdle of actually approaching them with a request to cooperate in filling out a lengthy protocol (see Appendix). How was I going to do this?

I tried three different approaches to data collection. One was a direct personal appeal to scientists with whom I, my brother or my wife, *etc.*, were acquainted. A second was to approach commercial organizations formally and ask if I might speak with staff. A third was to approach a professional organization (in criminal justice) and advertise for willing respondents. Only the first approach met with any success. It was a slow way to gather data but it proved to be the only way to elicit cooperation from this elite population.

I believe that my questionnaire generally worked well with the over 100 subjects who participated. But not everyone was pleased with every question. For some of my respondents, as for Zuckerman's Nobelists, precision of wording did prove to be an issue, at least in the case of the mail survey. One respondent, an eminent physicist (S042), refused to answer any questions that contained the words "political" or "politicized" or "politics" claiming he did not know what I meant. In a separate personal communication he stated that the ". . . questionnaire was seriously compromised by its emphasis on an ill-defined concept, *political*. I could answer your questions in many ways depending on how I chose to interpret this semantically loaded word."

My distinguished critic's comments were the sharpest criticisms I received. However, two other respondents to the mail survey indicated some concerns along similar lines. It is clear that the mail survey was probably the less desirable way to elicit meaningful data of the two data collection methods I used with the study populations. I believe, however, contrary to my critic's allegation, that much valuable data were collected at nominal cost through the mail survey. However, on some topics covered, it was appropriate to contact the respondents afterwards to clarify my mail survey questions and elicit further data from them by interview.

With a rather similar (if more "elite") respondent population to my own, Zuckerman (1979; originally 1977) used letters to introduce herself to Nobel laureates in science.

Initially, Zuckerman, tried "rather lengthy letters explaining the nature of the investigation" (1979:258). This was met with a decidedly unenthusiastic response from her busy respondents. Zuckerman's experience in this regard strongly influenced my decision to be brief and to the point in my approach to the potential respondents. Sometimes I wrote first but much more often I made my initial contact by telephone with prospective respondents. If I got a tentative consent, I then followed either with a very brief personalized handwritten note appended to the formal one page cover letter or to the instruction sheet in the questionnaire packet. Or, for interview subjects, I followed by bringing the questionnaire to the interview to demonstrate the self contained nature of the study. Zuckerman (1979:266; originally 1977) believed that the following facilitated her entree to Nobel laureates:

> The principal factors affecting receptivity . . . were the legitimacy of the interviewer's request, judged by her affiliations with Columbia University, the Social Science Research Council, and the National Science Foundation; the laureates' sense of obligation to other investigators; the self contained nature of the proposed interview; and not least . . . the sheer interest of the laureates in the subject of the interview.

My experience suggests that many of these same factors operated in my favor. Also, in many cases I had the benefit of personal acquaintance with close friends of my respondents.

The leitmotif in these observations is that when I sought the assistance of elite respondents I had to be "properly referred." This was evident when I tried simple advertisements for respondents in a professional society newsletter and in a large corporation research laboratory and met with no success. *Personal* appeals to my prospective respondents were necessary to obtain any offers of help. However, though time consuming to make, such appeals were successful in the great majority of instances tried.

My use of individual principal investigators as respondents is a methodological departure from Zeldenrust's work. Zeldenrust (1990:9) criticizes the "emphasis on individual scientists as units of investigation." He believes that it " . . . is likely to obscure higher-level organizational patterns and regularities, and brings about a focus on individual preoccupations with respect to problem solving tactics and careers."

I do not share Zeldenrust's concern. Properly done, an organizational analysis can rely on individual respondents without neglecting higher level organizational patterns which provide a context for helping to understand some of the individual behavior.

Caplow (1968)[10] addresses the concern raised by Zeldenrust in a rigorous fashion. He first acknowledges that "the private program of an organization member may never coincide exactly with the official program assigned to his position." (That the two do not completely coincide he explains as the result of the fact that (a) the "actions of men are always affected by a multitude of unconscious needs and by vestigial emotions developed earlier in life." Furthermore "transactions between an organization and its members embody a fundamental conflict of interest)." Nevertheless, Caplow (1968: 51-52) points out, "The inherent discrepancy between the collective interests of an organization and the personal interests of its members " is usually addressed by offering "higher status members" of the organization "more pay, perquisites and prestige in return for their organizational efforts" than others receive. "The intended result" is that in dealings between a superior and subordinates the superior "will identify the preservation of his own advantages with the maintenance of the status order and with conformity to the official program."

Caplow's views are echoed by other theorists, *e.g.,* Parsons, summarized in Landsberger, who says (1961:226)

> Parsons believes that managers in particular need to be collectivity oriented, both because they are responsible for the organization as a whole, and because employees' perception of the extent of management's collectivity orientation affects their readiness to obey.

Ironically, one of Parsons severest critics, C. Wright Mills, quotes Werner Sombart, Walter Rathenau, and others to support his contention that managers are in fact collectivity oriented to a degree that makes it seem as if organizations had motives of their own.

Landsberger[11] also makes the point that:

> "Fouriezos, Hutt, and Guetzkow, observing decision-making groups in government and industry, have indeed found a negative correlation between group productivity and average level of self-orientation."

My perspective is, thus, consistent with a considerable theoretical tradition, albeit one with limited empirical grounding. (Landsberger, in fact, says only that one study by Melville Dalton looks at the relationship of collectivity orientation of managers and organizational success and Dalton says the relationship is *negative*!)

Fieldwork

I did no fieldwork for my research on problem choice by scientists. Fieldwork is useful, even necessary, for certain kinds of studies and to obtain certain kinds of data. Indeed, Knorr-Cetina (1995:151-54) suggests that observations of interactions between scientists may be useful to study problem choice decisions (see her comments about "negotiation"). However, her review concedes that "most constructionist studies to date have failed to analyze the patterns and processes that turn an ordinary interaction into a 'negotiation.'" (1995:153).

Zeldenrust (1990:9) suggests, however, one reason why the fieldwork that Knorr-Cetina (1982) advocates is of limited value for the purpose of understanding research problem choice.

> [The drawback of] . . . the short time-scale of observation in . . . empirical studies, typically six months to a year . . . [is that] in a period of that length an observer is likely to detect a great deal of 'noise,' with only an occasional choice incident of major significance (which may then not be interpreted in the context of previous significant choices, or compared to other such events).

Zeldenrust is particularly concerned that social constructivists have emphasized the "inherent chaos" of investigative activity. They would be especially misled in this direction for three reasons. First, the laboratory studies on which they base their theoretical ideas are too limited in time duration to spot the underlying regularities. Second, the focus on individuals rather at the organization level entails dangers of its own and third, the individual scientist in a research team may be confronted more often than the organization with what appear to be changing sets of choices. The organization, not concerned with details, sees more uniformity in its overall strategy and behavior.[12]

These changing sets of choices, however, may come about through organizational level efforts at environmental control that the field worker's focus on individual researchers can miss. This deeper stability of the organization environment cannot be understood by field work observation alone. It requires the field worker to focus on higher levels of organization and data relevant to each.

To assess that organization environment the researchers may need to look at archival materials, interview executives, *etc.* (though Zeldenrust himself does *not* address this methodological issue).

Unlike social constructivism, the rationalist perspective represented by L-Z-F grasps the importance of organizational and higher level factors in problem choice. This is because the rationalist sociologist would begin with the premise that research problem choice is a more orderly process than field workers find it to be. Therefore, to the rationalist it should be possible to (1) ask scientists what they are currently researching and why they chose it and (2) ask them what their next research project is likely to be. Then, unless something unexpected happens, most scientists can be asked sometime later whether they did what they told the rationalist sociologist they were planning to do and, probably, they would say that they did.

While I began as a rationalist applying survey methods, I had to rethink my rationalist point of view based on my survey data. My data on the topic "Future Directions" Q.75a-c are apposite here. Given my initial rationalist views, I fully expected my scientist respondents to be able to provide "the title and a brief description of your next project." I was indeed surprised by the answers I received. As reported elsewhere in this study, a large number of respondents did not know what their next project would be (see Chapter 5). Some could not say how they would go about choosing the next project, and in cases where they could, they might point out that someone would bring them their next assignment. Perhaps the respondents would tinker with the assigned problem to make it scientifically "doable." However, they clearly were not doing their problem selecting from a collection of known problems in their discipline which, as a rationalist, I had supposed.

Anonymity Guarantee

Sociologists have a special responsibility not to expose their respondents to unwanted attention as a result of the sociologists' behavior. Junker (1972:96) says:

> It is abundantly clear that not only the gathering of relevant data, but also their analysis and presentation, pose difficulties. Although the function of sociology is not to expose, yet it proposes a type of analysis which requires the use of facts often hidden from the public and sometimes even from the conscious thinking of the individual actor. Expose must be used insofar as it is necessary to analysis, but . . . the sources cannot be identified.

A study of problem choice in science may not seem at first glance to be especially controversial or to engender anxieties in respondents that they are going to suffer embarrassment or worse if their identities are revealed. However, even this study probed some areas that respondents might be reluctant to discuss openly unless their identities were not revealed, *e.g.*, interpersonal competition, conflict, and politics in science. Anonymity was given to all respondents. No answers are attributed to people by name. In this essay no other respondent was told what a respondent said and only a few respondents even knew who *any* of the other potential respondents were going to be. These few who do only knew about respondents they had asked to participate. But only the author of the essay knows the identities of all of the respondents.

ENDNOTES TO CHAPTER FOUR

[1]On 'social circles' see Kadushin (1966; 1968). A brief review of Kadushin's findings is found in Crane (1977:174). This question was inadvertently omitted from an early distribution of the questionnaire. Though theoretical grounds for imaging limited autonomy of scientists already exist in American sociology, until my study, no one has actually asked American scientists, including academicians, government scientists, *etc.*, if they feel they have freedom to choose the problems they want to study.

[2]Zeldenrust's lack of interest in this variable, I presume, is a reflection of the limited number of case histories Zeldenrust had for his study. Perhaps the organization cultures were not sufficiently varied for him to notice them.

[3]Shenhav's challenge to the assumption of scientist autonomy may not have had the impact I think it deserves because his data were responses of *Israeli* scientists in the academic environment. Israel throughout its history has been legally in a state of war with some or all of its neighbors. It remains (as of 2004) legally at war with one of its principal neighbors, Syria, and also with Lebanon. Fighting has erupted at least four times since the country achieved independence: 1948 when the state of Israel was declared; 1956 during the Suez Canal crisis with Egypt; 1967 when Israel captured Jerusalem and the Golan Heights which it still retains, and 1973 when Egypt briefly pushed Israel back from the shores of the Suez Canal in the Yom Kippur War. (Subsequently the two nations signed a peace treaty.)

Given its unusual situation, the fact that Israel's academicians do not feel they have control over their problems is *not* indicative that the rationalist hypothesis is wrong. Israel may simply be an anomalous case where overwhelming needs of national defense limit academic freedom and scientist autonomy.

Shenhav's (1985) contribution, however, adds to a rising tide of sociological theory that casts doubt about the assumption that scientists have professional control of their problems. European writers, in particular Knorr-Cetina (1995), Nederhof and Rip (Ms., n.d.), and Zeldenrust (1990), have pointed out that today science is *collectivized*, that scientists are usually employees, and thus professional autonomy in problem choice is really not a tenable assumption for a large percentage of scientists.

The point of the European writers about science is hardly novel. As far back as 1960, Talcott Parsons (1960:62-63) made much the same point about professionals in bureaucracy when he observed that:

> . . . decisions made in the management system control the operations of the technical system. This is certainly true for such matters as the broad technical task which is to be performed in the technical system - the scale of operations, employment and purchasing policy, *etc*. But, as in other cases of functional differentiation, this is by no means simply a one-way relation, for managerial personnel usually are only partially competent to plan and supervise the execution of the technical operations. The manager presents specifications to the technical subsystem, but vice versa, the technical people present "needs" which constitute specifications to the management; on various bases the technical people are closest to the operating problems and know what is needed. Perhaps the most important of these bases is the technical *professional* competence of higher personnel in technical systems, a professional competence not often shared by the administrative personnel who - in the line sense - are the organizational superiors of the technicians.

[4]In my study, I tested the assumption of scientist autonomy using as indicators:

- a Did the scientists claim to have great freedom to choose whatever problem she wanted to work on?
- b Did a sponsor come and tell her to do a specific piece of research?
- c Did scientists report that the organization they work for wants them to work on problems brought to the scientists by their supervisors?

The second indicator is essentially the same one used by Shenhav (1985) in his study.

The point to remember is that despite ample grounds in literature for supposing that professional autonomy of scientists is limited, refuting the rationalist hypothesis requires more than refuting this assumption of autonomy.

[5]Converse (1986:45) stresses the desirability of *multiple* indicators:

"Investigators wisely seek multiple and scaled measures but there is no guarantee they will remain scaled. They may be scattered by the words of change into component or single items that require individual interpretation."

She concedes that "single questions survive, too, for the simple reason that one can never, in a single survey, incorporate multiple measures of *everything*." However, she emphasizes that □Multiple measures are the strategy of choice. Relying on single questions make it difficult to uncover complexity. Using multiple indicators makes it easier to discover where or how our understanding of the world is inadequate."

[6]Because this indicator was added to the questionnaire after the questionnaire was initially developed, there are fewer respondents for whom actual data are available (N = 62). However, most respondents answered this question. Statistical analyses of the missing cases were done to determine if systematic bias was present; it was not.

[7]Kirk and Miller (1986:22) define "apparent validity" by example. They say that "In the best of worlds, a measuring instrument is so closely linked to the phenomena under observation that it is 'obviously' providing valid data." Examples are "Formal examinations of competence and achievement (academic, civil service, professional tests)." They add, however, that " the validity of measurements is too seldom evident on the face of things. Conclusions of apparent validity are not entirely out of order, but they can be illusory."

[8]A lengthy discussion of the considerations that led to this choice is found in the Technical Appendix to Chapter Three.

[9]Hall, Oswald, 1944 "The Informal Organization of Medical Practice in an American City (Ph.D. diss. Dept. of Sociology, University of Chicago, quoted in Junker, B. (1960) p. 95.

[10]Caplow (1968), *Two Against One* Englewood Cliffs, Prentiss-Hall. Caplow echoes a view Talcott Parsons has advocated.

[11]See Fouriezos, Nicholas T., Max L. Hutt, and Harold Guetzkow (1953) "Self Oriented Needs in Discussion Groups," pp. 354-60 in Darwin Cartwright and Alvin Zander (eds.) *Group Dynamics: Research and Theory* Evanston, Ill.: McGraw, Peterson and Co.

[12]Zeldenrust(1990: 9) maintains that: ... the emphasis [in

constructivist studies] on individual scientists as units of investigation is likely to obscure higher-level organizational patterns and regularities,…." Furthermore, it can "bring about a focus on individuals preoccupations…."

CHAPTER FIVE: DO SCIENTISTS CHOOSE PROBLEMS RATIONALLY?

Comparing The "Garbage Can" Model and Other Perspectives

> It is sometimes possible for the data to be so strong, and the connections between the data and the rival models to be so obvious that ... whatever their original personal, professional, or social interests, it is possible for the vast majority of a profession to be left with no satisfactory option but to accept the new models as the best available representations of the world.
>
> Ronald N. Giere
> *Explaining Science* (1988:227)

INTRODUCTION

Up to this point, I have been setting the stage. In Chapter One, I stated my intent in this study to show why there is a connection between gender differences in productivity and the problem choice process in science. To do this I would need to evaluate the various current models of problem choice against the data I had gathered and also explain why there should be gender differences in problem choice.

In Chapter Two, I set forth my general strategy for evaluating the various models of problem choice and for explaining gender differences. I also stated hypotheses that I expected would help identify which model of problem choice would be the best candidate to be a foundation for an empirical theory. And I made clear my preference for the "garbage can" model as opposed to either the rationalist hypothesis or its principal rival, the social constructivist tradition.

My strategy for demonstrating the superiority of the "garbage can" model is to:

1. Show the patterned differences in problem choice criteria between the genders among more autonomous scholars,
2. Show that the impact of organization influence can be positive as well as negative and
3. Show that research problems can come from anywhere.

If I can demonstrate patterned gender differences in problem choice criteria, I can cast doubt on the social constructivist argument about the chaotic nature of the problem choice process. The social constructivists, while accepting that personal characteristics play a role in problem choice, do not believe that *patterned* differences between the genders exist. If I can demonstrate that the gender patterned differences will be greater among the more autonomous researchers than among less autonomous researchers, I can also cast doubt on the rationalist hypothesis. The rationalists permit gender differences but see these as resulting from gender differences in professional autonomy.

In regard to the second point, the rationalists accept that organizational influences on problem choice can exist. However, they see them as negative since they assume that researchers know what the most important problems are and are drawn to solving them.

Regarding the final point, my showing that problems can come from anywhere would throw doubt on the central tenet of the social constructivists that problems are locally originated. Interestingly, Zeldenrust (1990), who is critical of the social constructivists on a number of points, appears to accept local origination of problems it as an empirically correct observation by the social constructivists.

It is important to point out that not all social constructivists insist that problems are locally originated. Since Knorr-Cetina, a leading social constructivist, allows for multiple discoveries, she implicitly permits problems to be derived from theories. This is evident because she accepts that similar equipment in different laboratories can lead to similar results. The rationalists do not formally take a view on where problems come from but seem to lean to the view that they are derived from theories with few exceptions.

Before I begin the presentation of my findings, it is important to reiterate that the other two rival perspectives continue to enjoy support for good reason.

For example, though it has been fiercely attacked in the literature (Knorr-Cetina [1981; 1983; 1995], Zeldenrust (1990), and even its assumptions questioned (Shenhav [1985], Zeldenrust [1990]), the rationalist hypothesis continues to be the standard hypothesis of scientist problem choice because of its clarity and its intuitive plausibility. It makes sense to expect that scientists will be drawn to problems consistent with their theoretical and methodological commitments, as the rationalist hypothesis claims. Theories, after all, are the conceptual tools of disciplines and researchers honor theories when they use them in their work.

Most important, there is no compelling evidence in the literature that the rationalist hypothesis is incorrect in its predictions. There are some studies addressing "problem specialty choice," (*e.g.*, Gieryn [1978], who champions the rationalist hypothesis), that are consistent with the rationalist perspective. There are also some small case studies of laboratories reviewed in Knorr-Cetina (1995), which indicate that the hypothesis does not fit their findings. Yet, none of these studies constitute a test of the rationalist hypothesis. Indeed, despite all of the criticism from historians (Westfall [1977], organization analysts (Shenhav [1985]); Zeldenrust [1990]); sociologically oriented natural scientists (Ziman [1980]); and sociologists of science (Knorr-Cetina [1982]; Latour [1979]; Collins [1982]); there is not a single study that tests the hypothesis. (Shenhav [1985] hardly addresses the hypothesis at all though he marshals evidence casting doubt on the plausibility of the assumption of scientist autonomy. That assumption, of course, is generally considered central to the rationalist perspective).

The key assertions of the social constructivist hypothesis are well grounded in case studies of laboratories (Latour and Woolgar[1979]; Knorr-Cetina (1981]). For example, it is clear that medical scientists draw heavily on prior findings in deciding their next research projects. And these medical studies do not draw upon a comprehensive paradigm the way studies in atomic physics and some other specialties do. If nothing else, the "laboratory studies" of the social constructivists make a cogent case that the fundamental assertions of rationalists -- problems are chosen based on theory commitments by autonomous scientists -- are not universally valid.

Thus, there is a lot of smoke. Is there a fire? It is time to scrutinize the three perspectives and ask hard questions: (1) can they be falsified? (2) is it possible to devise a reasonable, fair test of these perspectives and determine if one is clearly better than the others? In this chapter, I try to answer these questions in detail.

Let me begin with the rationalist hypothesis. What precisely is a fair test of the hypothesis? The hypothesis is subtle and it must be understood in all its subtlety. First, it is based on an assumption that scientists have autonomy in their problem choice.

Second, the hypothesis says that the scientists use certain criteria in their problem choice: (1) the problem must be important from the standpoint of the paradigm in the discipline except occasionally and/or must be regarded as intellectually important by the leaders of the discipline. Above all, it must be *doable*. (In the Lederberg-Zuckerman-Fisher version, the problem chosen must be feasible from the standpoint of organizational, politico-legal, and technical criteria. This incorporates Fujimura's insight about constructing doable problems through "articulation work" [1987]. Zeldenrust [1990] opposes the Fujimura viewpoint to the rationalist perspective but I do not see them as inconsistent. Against this rationalist hypothesis stands the contrary view, called the social constructivist perspective, identified with such scholars as Knorr-Cetina (1981;1983) and Latour (1979).

As summarized by Zuckerman (1988:555), these scholars maintain that:

(a) problems are locally originated and

(b) problems are largely or entirely contingent; they result from scientist decisions that are "opportunistic," *not* "well-planned" and *not* "rule governed."

Finally, there is a third model (Zeldenrust 1990). It contains some features of the social constructivist tradition and a few aspects that seem to be consistent with the rationalist viewpoint. This "garbage can" model, although described in Zeldenrust's work, cannot be regarded as having been tested since the data were case studies based on archival records. Based on literature addressed in this study however, I find the model more persuasive than either of the other two models. Still, it is necessary to examine all the models against data gathered to assess their ability to explain problem choice. And now that the stage is set, it is time to turn to the evidence I have gathered from my survey to see which model - the "garbage can" model or an alternative model - explains better how research problem choices are made.

FINDINGS

Autonomy

I will begin with findings bearing on autonomy of scientists in problem choice. The reason for this is simple: first, the main hypothesis of the rationalist model presumes scientists are free to choose whereas the "garbage can" model and the social constructivist models do not. Second, regardless of which model is more correct on this count, I am concerned with the broader question of how problem choices are made in scientific research organizations. It is important, therefore, to establish at the outset whether scientists themselves make these choices free of interference. Based on my review of the pertinent research (*e.g.*, Shenhav [1985], Westfall [1977]) I do not think they are free now and I do not think they were free in the past. I am not referring to "bottle washers" in making this assertion. The literature suggests that even top scientists share responsibility for problem choices with nonscientists (political leaders, corporate managers, *etc.*). In other words, scientists are usually participants in problem choice but not necessarily the decision makers. Therefore, it would be difficult to find scientists who are entirely free of nonscientist influences in choice decisions.

More precisely, I would argue, from a role theory perspective, that the *role* of scientist is to *suggest* problems. A scientist can play other roles as well. For example, the same person can be an "entrepreneur" as well as "scientist."

The entrepreneurial role, or the "champion" role, performs certain integrative functions with institutions of society; specifically, it links the role of scientists to the larger social system's "demands" for the work of scientists.

Although scientists *may* embody in themselves both the *role* of scientist and one or more other roles central to the choice process, they cannot be presumed to embody the role of entrepreneurs. Thus, it is more correct to speak about problem choice in science rather than about a scientist's problem choice. (Zeldenrust makes this point also by indicating that only partial linkages are possible until "demand" and "problem" and "constraint" are all linked.)

The involvement of other people, often themselves nonscientists, in the problem choice process is acknowledged by the social constructivist scholar Knorr-Cetina (1982: quoted in Zuckerman [1988;539]). However, Zuckerman counters that "as organized for some time now, prospective scientific work in the form of research proposals is assessed primarily by peers drawn from the same specialty, or inter-specialty . . . ."

Which view is correct? Are scientists freely choosing problems or are organizations choosing problems? In regard to this question, I investigated the following hypotheses:

Hyp: 3.1 **The proportion of scientists who regard themselves as having great freedom to choose their problems is greater than ninety (or ninety-five) percent of the total as indicated by:**

Hyp: 3.1.1a: **The proportion of respondents stating "I had a great deal of freedom to choose problems I wanted to work on" is greater than ninety (ninety-five) percent.**

Hyp: 3.1.1b **The proportion of respondents stating "a sponsor told me specific research to do" is less than ten percent.**

Hyp: 3.1.1c **The proportion of respondents stating that organizations want them to investigate problems brought to them by organization superiors is less than ten percent.**

Table 5.1 shows that 78% of the 105 cases for whom I had adequate data claimed to have freedom to choose problems.

Table 5.1: Indicators of Scientist Autonomy in Problem Choice by Gender

	Gender		
	Men	Women	Total
Can Choose Own Problems	87% (55)	66% (27)	78% (82)
	p = .015 (Fisher's Exact Test)		
	N = 105		

Sponsor Selected Problem	36% (23)	38% (15)	37% (38)
	p > ..517 (Fisher's Exact Test)		
	N = 104		

Organization Wants To Pick Problems	47% (29)	53% (16)	49% (45)
	p > .357 (Fisher's exact test)		
	N = 92		

Though researcher respondents in my study overwhelmingly claimed that they had considerable freedom to select problems, it is possible that they had been constrained in some fashion in performing their latest research, as Nederhof and Rip (n. d.) argued in their analysis of Dutch biotechnology teams.

Therefore, I asked two additional questions: (a) "Did a sponsor tell you the specific research to do" in your latest research and (b) Does your organization want you to investigate problems brought to you by organization superiors?

For the latter question, I had data from 92 respondents, 45 of whom (49%) stated that indeed they worked for organizations where this was true. However, this is not convincing evidence that scientists lack autonomy since (in theory, at least) all the superiors could be themselves scientists.

Furthermore, in the case of the second question, although 38 respondents, more than a third (37%) of the 104 respondents from whom I had valid responses, claimed that the sponsor told them what to do, it is quite possible that this was only after the researcher had proposed the problem and the sponsor responded with an assent. In other words, I have no incontrovertible evidence in these particular data that the assumption of scientist autonomy is unrealistic.

However, I also looked at whether, as predicted by Hyp.3.1a, there were patterned differences among scientists in their degree of autonomy. Specifically, I wanted to know if men and women scientists were likely to enjoy different degrees of such freedom to choose. As Table 5.1 shows, indeed there are such gender based differences in my sample; 66% of the women (N=27) versus 86% (N=55) of the men claimed such freedom among the 105 cases for whom I had usable data.

The difference between men and women may be traceable to

different experiences men and women respondents had with institutional controls where they worked. Among the 94 respondents for whom I had data, 39% of the women (N=14) versus 16% of the men (N=9) asserted that they had to alter or forego a study at some point in their careers because of institutional controls. (See Table 5.20 *infra, p.205.*).

Rationalists might point to this finding as indicating that women scientists may be more constrained than men by society. The rationalists would not regard the finding of gender differences as having negative implications for their hypothesis, which only argues that *if* scientists are free to choose, they will select the most significant problems they can. Furthermore, some data gathered for this study seem consistent with the rationalist hypothesis. For example, when I looked at the relationship of "Freedom to choose own problem" by "novel analytical approach)," I found that of 104 respondents those who claimed the most freedom were more likely (54%, N=44) than those who claimed less freedom (35%, N=8) to say that in their more recent project, they considered using a "novel analytical approach" to the data (Table 5.2B).

Table 5.2A: New Research Approach by Sponsor Selected Problem

	Sponsor Selected Problem			
New Research Approach	Yes		No	
	Number	Percent	Number	Percent
Yes	14	36%	37	57%
	p = .039 (Fisher's exact test)			
	38		65	

Table 5.2B: New Research Approach by Can Choose Own Problems

	Can Choose Own Problems			
New Research Approach	Yes		No	
	Number	Percent	Number	Percent
Yes	44	54%	8	35%
	p = .078 (Fisher's exact test)			
	81		23	

Given this finding in Tables 5.2A and 5.2B, I decided to look more closely at the rationalist contention about the consequences of women respondents differing in their reported freedom to choose problems. I reasoned that if the rationalists were correct, gender based differences in problem choices should disappear if women were as free as men to select their research problems.

On the other hand, however, suppose that gender based differences were actually *greater* among those free to choose their problems. This surely would be a difficult challenge for the rationalist to explain away. They would be forced to concede that scientists do *not* have perfect information about the significance of problems. And that is exactly what the "garbage can" model would expect.

To test the divergent predictions of the "garbage can" model and the rationalist hypothesis, I compared men and women scientists who claim to be free to choose their problems (*i.e.,* have relatively high degrees of perceived professional autonomy). Given the small numbers of respondents, especially female respondents, in this exploratory study, *very* large differences in the dependent variate parameter values for each gender would be needed to achieve statistical significance. Thus, the odds are stacked in favor of the rationalist perspective being correct since it predicts that under conditions of free choice gender based differences should disappear.

What did the data show? Tables 5.3Aand 5.3B present findings pertinent to this issue. In Table 5.3A, I present several criteria of problem choice cross-tabulated by gender, given that the respondents claim that they "*had a great deal of freedom to choose problems*" they wanted to work on. For example, in regard to whether the researcher considered if she would have access to special equipment not always available to her, 49% (N=27) of the men versus only 26% (N=7) of the women claimed that this was a criterion. In the case of whether the problem was socially or politically significant, 83% of the women (N=19) versus 53% of the men (N=29) claimed that this was a criterion.

I also found significant differences between men and women on other criteria, lending support to the "garbage can" model and undermining the plausibility of the rationalist hypothesis.

In sum, what conclusions can we draw from the evidence about the utility of current problem choice models? Three main findings bear on the models of problem choice. First, despite the fact that the study sample is biased *deliberately* in favor of experienced senior scientists, just 79% claimed that they had freedom to choose their problems. More tellingly, 61% of those for whom I had valid responses claimed that "the organization wants me to work on problems brought to me by organization superiors." These results echo those of previous studies, *e.g.,* survey research (Shenhav [1985]); (Nederhof and Rip [Ms. n.d.]) and historical research (Westfall [1977]) which strongly question the autonomy of scientists in problem choice.

Table 5.3A: Past and Present Criteria of Problem Choice by Gender of Scientists Among Highly Autonomous Scientists*

	Gender			
	Women		Men	
	Percent	Number	Percent	Number
Public issues usually important in my problem choice	83%	19	53%	29
	p = .011 (Fisher's exact test)			
	23			55
Wanted to beat a competitor	23%	6	49%	26
	p = .023 (Fisher's exact test)			
	26			53
Access to special equipment important now	26%	7	49%	27
	p = .038 (Fisher's exact test)			
	27			55
New research approach important now	41%	11	61%	33
	p = .067 (Fisher's exact test)			
	27			54

* *Measured by whether they stated that they can choose own problems*

Table 5.3B: Past and Present Criteria of Problem Choice by Gender of Scientists Among Highly Autonomous Scientists*

	Gender			
	Women		Men	
	Percent	Number	Percent	Number
Prefer problems usually related to public concerns**	86%	19	59%	24
	p=.021 (Fisher's exact test)			
	22			41
Desire to do before rival did it a criterion in present research	13%	3	33%	13
	p=.058 (Fisher's exact test)			
	23			40

Desire to do it before someone else did it a criterion in past	16%	4	53%	21
	p=.003 (Fisher's exact test)			
		25		40
Access to special equipment a criterion in current research	20%	5	49%	20
	p<.05 (Fisher's exact test)			
		25		41
Possibility of using novel analytical technique in current research a criterion	36%	9	70%	28
	p=.007 (Fisher's exact test)			
		25		40

**Measured by whether they stated that the sponsor did NOT select the problem chosen*

***I also asked a related question (Q21p) and got similar results: e.g., 76% (N=19) of females and 51% (N=21) of males among those claiming that a sponsor did not tell them what research to do asserted that a problem's social or political significance was a criterion in choosing the problem*

There is still the point made by Zuckerman to contend with, namely that the research proposals of scientists are reviewed by their peers (Zuckerman, 1988). My answer to her is that Zuckerman is referring to a small fraction of the scientific research community. Most industrial research probably does not occur in the way Zuckerman describes. The same may well apply to much government research in its own laboratories. Research proposals are usually employed to link government and private funding sources with university based or other outside researchers in the United States.

The second main point is that the results reported here do more than merely add to the accumulating literature questioning the plausibility of scientist autonomy in problem choice. My findings point to a *cogent reason why* scientists cannot be presumed to be making problem choices free of nonscientist influence. Scientists do not necessarily know the comparative intellectual significance of problems they may have an opportunity to work on. In the absence of such knowledge, they are perfectly willing to look to nonscientists for guidance on what research would be desirable from their standpoint. The finding that many scientists seek problems related to public controversies illustrates this point.

The third point I wish to make in this summary is that women enjoy even less autonomy in problem choice than men. Proportionately fewer women reported having freedom to choose their problems (see Table 5.1). There is suggestive evidence that they experience more requests than men to stop research for other than technical reasons (39% versus 22% P.=.096, Fisher's Exact one-sided test) and Table 5.20 shows that women more often had to forego or alter a study because of institutional controls.

The fourth point is that among more autonomous scientists there are greater discrepancies in preferred problems between men and women than among less autonomous scientists. This is exactly the opposite of what the rationalists would predict.

It is possible to argue that these results were found in just one small data set; furthermore, that the indicators of autonomy are open to question and therefore that these results do not clearly establish the superiority of one of the models. Yet, in the context of the results of many other studies, my findings should cause rationalists to wonder how plausible the assumption of "autonomy" is. Not only are there numerous studies casting doubt on the assumption's plausibility, there is now also a reasonable explanation for why autonomy is not something that should be presumed beforehand: when scientists do not know which problems are more important, (do not have perfect information on problem significance), they will be more inclined to seek guidance from various sources, including nonscientists, in deciding what to study.

This leads us to the next issue, where do scientists find their problems? Perhaps under conditions of imperfect information about problem significance, Knorr-Cetina is correct that nonscientists have a role in problem choice. This, of course, raises an interesting question: if nonscientists can influence problem choices in science, what are the implications of this finding for Kuhn's belief (1969), which Zuckerman seems to approve, that problems emerge from dominant paradigms? Does nonscientist influence in the choice process bolster the social constructivist argument that problems originate locally? These questions will be addressed in the next section of this chapter.

Theoretical Derivation of Problems versus Local Origination

It is quite easy to define the views of prominent social constructivist scholars, e.g. Knorr-Cetina (1982), H. Collins, (1982) Latour and Woolgar, (1979), on where problems are found: problems are *locally originated.* Even Zeldenrust (1990), a critic of the social constructivists on several grounds, remarks that problems of the scientists in his study were in fact locally originated.

It is important to mention here that even while maintaining that problem choice is local, Knorr-Cetina (1981:6), in trying to account for "multiple discoveries," implicitly argued for a role for theories commonly shared by numerous independent scientists. This is plain in her references to "similar educations" and equipment commonly found in the many research centers pursuing research on similar problems.

Unlike the social constructivist writers, the rationalists do not clearly spell out their position on where problems are found. However, Thomas Kuhn (1969; originally 1962) who influenced many writers in the rationalist tradition, believed that theories "define" problems (see endnote #8 at the end of this Chapter). There are also many hints in Zuckerman's writings that she shares Kuhn's position that problems come from theories *e.g.* when she points to a "shared knowledge base" and "similar agendas" in discussing why multiple discoveries occur in science.

Whereas the social constructivists argue for a local origin of problem, and the rationalists seem to lean towards theoretical derivation, the "garbage can" model quite clearly takes no position at all on where problems are found. (The reader should clearly see that Zeldenrust's "problems" are simply March's *et al.* "solutions" in a new context. Scientists are offering "problems" [as solutions] to meet the demand they perceive or in anticipation of demand they will seek [by "unblackboxing"]. This "demand," in the present context, is March's "problem" – albeit combined with March's resources looking for work. More precisely, a main point of the "garbage can" model is that problems can come from *anywhere* and can be *inserted* at anytime and *withdrawn* at *anytime).*

In conclusion, if my data show that chosen problems are locally originated (*i.e.*, result from findings of prior studies at the research site) then this would enhance the social constructivist position. If problems are almost always theoretically derived, and/or are investigated simultaneously at more than one research site, this should diminish the plausibility of the local origination hypothesis in the social constructivist perspective while enhancing the rationalist hypothesis.

With this in mind, I investigated the following general hypothesis:

Hyp. 3.2: **Scientist respondents will not always respond either that problems are locally derived or theory derived.**

The specific hypotheses were:

Hyp. 4.1.1a: **Scientists will report that they obtain the problems they will research by comparing problems with one another.**

Hyp. 4.2: **Scientists will report that the latest problem they chose was suggested by "theory in my own field" or "after respected colleagues urged work on them."**

Table 5.4 shows that, for my respondents, both in the case of their current research problem and their next problem (when known) available theory suggested the problem about half of the time (52% and 49%). If problem choice is local, this many respondents finding problems derived from theory appears to make no sense. But these results also do not justify a belief that problems are *theory* derived.

Table 5.4: Comparison of Different Disciplines on Indicators of Rationalist Approach to Problem Choice

	YES	NO	N
Respondent was able to give title/describe *next* project	31%	69%	62
a. Biologists/med. Only (N=15)	27%	73%	
(Next) Problem tests a theory important in the discipline	52%	48%	101
a. Biologists/med. Only (N=35)**	34%	66%	
b. All others (N=66)	62%	38%	
(Current) Problem was suggested by theory in my own field	49%	51%	104
a. Biologists/med. Only (N=34)	50%	50%	
b. All others (N=70)	49%	51%	
Respected colleagues urged work on (next) problem	47%	53%	93
a. Biologists/med. Only (N=31)*	58%	42%	
b. All others* (N=62)	42%	58%	

Current problem was considered important by renowned researchers	48%	52%	105
a. Biologists (N=34)	53%	47%	
b. All others (N=71)	45%	55%	
Compared and then chose current problem	49%	51%	105
a. Biology/medicine (N=35)*	66%	34%	
b. All others (N=70)*	41%	59%	
* ***Significant at the .05 level (Fisher's exact test)***			
** ***Significant at the .01 level (Fisher's exact test)***			

Recall that Zuckerman hypothesized that problem choices are made from the stack of previously identified problems. But these do not have to be problems in the published literature. Maybe through informal communication of some sort, influential colleagues brought these problems to the researchers' attention. Perhaps the process of problem choice, in other words, resembles the process of opinion formation (see Elihu Katz and P.F. Lazarsfeld, *Personal Influence*, Glencoe, IL: Free Press, 1955) in some respects. For example, we might suppose that certain people evaluate the voluminous professional literature and decide what is worthwhile. Then others, based on these assessments, choose problems suggested by the *recommended* literature. Or perhaps scientists differentiate between (a) journal publication and (b) more informal communications, *e.g.,* e-mail or other personal communications, advance drafts of articles submitted for publication, *etc.* The former is "literature" while the latter is more generally considered a suggestion from renowned researchers.

The bottom line is that I needed to look further in order to cover these possibilities. I began by taking a broader view of theory as a source of problems to indicate rationalist problem choice, For instance, I also looked at suggestions of renowned researchers. (See Table 5.4) The results were about the same as with theory.

Neither in regard to the "next" problem the respondents planned to do (47%) nor in regard to current problem (48%) did respected colleagues influence more than half of the respondents. Thus, neither literature nor respected colleagues were sources of influence on problem choice for about half (or more) of the respondents.

I then asked a further question, Did the researchers weigh, or compare, the chosen problem against any others that they identified? Of the 105 respondents for whom I have valid responses, 51% (N=53) said they chose their current problem after comparing it with another. The biologists and medical researchers (70%) were more likely than others (42%) to compare problems before choosing one that was preferable. (In some disciplines, however, this comparison of problems is apparently so uncommon that one respondent commented that asking her if she chose her research problems after comparing them to others was a "strange" question!)

Because some of the findings challenge the social constructivist position on local origin while others support their positions, it seems to me that the main beneficiary of these findings is the "garbage can" model. (The rationalists hardly benefit because to the extent that rationalists share Kuhn's position that "theories define problems," a substantial number of the respondents were giving answers that seem absurd to a rationalist, though "obvious" to a constructivist).

Yet, we are still a long way from rejecting the rationalist hypothesis in favor of the "garbage can" model's position of neutrality on where problems are found.

A more sensitive test is necessary to assess the rationalist perspective and for this I needed to construct an "index of Kuhnian method." I reasoned that, although no single indicator was especially good, an index would demonstrate the correctness of the rationalist hypothesis. Accordingly, I constructed an index based on three key indicators:[1]

(a) whether the problem chosen "was suggested by theory in my own field,"

(b) whether "I compared several ideas for research topics with each other" and

(c) "whether the problem has been assessed as important by renowned researchers in the field.

The Index of Kuhnianchoice process showed that 77% (N=80) of the 104 respondents went about their problem selection as the Lederberg-Zuckerman-Fisher version of the rationalist perspective would predict.[2]

The next step was to evaluate this result using the "goodness of fit" chi-square criterion. I chose two strong criteria. One was a ninety percent criterion and the other was an even stronger ninety-five percent criterion. In other words, if the researchers were indeed making selections as the (Lederberg-Zuckerman-Fisher version of the) rationalist hypothesis would predict, then, for example, by the ninety percent criterion there would be no statistically significant difference between the 80 observed "rationalists" and the 94 expected rationalist cases out of 104 respondents. Similarly, if the 95 percent criterion were met, there would be no statistical difference between 80 *found* and 99 *expected* cases.

As shown in the Technical Appendix, by the ninety percent criterion, the null hypothesis -- in this case the *rationalist hypothesis* -- is accepted. However, by the more stringent ninety-five percent criterion, it is rejected. I am more impressed by this latter finding than by the finding of support for the null hypothesis at the 90% criterion level. I believe that far fewer of the scientists than might have been expected by Kuhn obtained their problems from received theory. However, the rationalists might not agree with my interpretation seeing in the less stringent test results within this small data set adequate support for their views.

This is a good point to recapitulate. First, I questioned the assumption of researcher autonomy based on my findings about gender differences on problem choice criteria. (Furthermore, Nederhof and Rip (n.d.)using a random sample survey also cast doubt on the plausibility of this assumption.)

Now I have presented data that do not strongly support the case that problem choices almost always are theoretically derived. On balance, only the "garbage can" model seems to benefit from these findings on where problems are found. The case for the social constructivist model is undermined by the findings that scientific problem choices are not often truly local choices while the case for the rationalist hypothesis is not helped by my finding that theoretical *gaps*, an absence of theory on some phenomenon piquing the interest of the scientists, can lead to them offering a problem for research. Zuckerman (1978)is quite explicit that problems are picked (by scientists free to choose) based on theoretical commitments except in a few cases. However, it appears to happen far more often than she imagined that scientists identify a *theoretical gap* and try to close it.

It is now possible to see Holton's remark that "the discovery of new areas of ignorance excites scientists to vigorous activity" (see Zuckerman [1978:85]) in a new light. Considered against the evidence reported above and in Chapter Six (see fn. 1), it is clear that Holton is correct regarding what occurs in disciplines with strong paradigms; there, "new areas of ignorance" *created* by the occasional addition to the theoretical literature, (presumably, in most cases, contributed by one of the acknowledged leaders of the discipline) have excited such scientific activity. However, particularly in disciplines with weak paradigms, my results suggest that spotting phenomena unaccounted for by any theory also excites new research

Competition for Priority: Evidence of "Shared Problems"

1. Comparison of Problems

What are perhaps most interesting in the data on where problems are found are their implications for the rationalist view that problems are "shared." (Recall that one key reason why rationalists expect problems are shared is that they expect most problems to be derived from theory, -- a view clearly influenced by Thomas Kuhn (Kuhn[1969; originally 1962]) judging by the liberal quoting of his work in an approving tone by such scholars as Zuckerman (1978).

It is not essential that problems be theory derived to be shared but it seems less likely to me that problems not theory derived are shared. Therefore, if a substantial number of scientists are denying that their problems are theory derived, then this throws some doubt on the argument put forth by Gieryn (1979) that problems are "shared."[3]

I decided to test the idea that there might be differences in proportions of "shared" problems between those derived from theory and those not derived from theory. For this analysis, I used a proxy measure for "shared" problems, "did the scientist *compare* the problem chosen with other problems before selecting it?" Two-thirds (67%) of respondents claiming the problem was derived from theory also said it was selected after comparing it to other problems, versus 36% among those who claimed the problem was not suggested by theory.(see Table 5.5 below). This bolsters my conviction that the rationalist assumption of "shared" problems is questionable.

If it cannot be assumed *a priori* that problems are shared, then the most interesting prediction of the rationalist hypothesis (which even many nonrationalist scholars such as Mulkay [1991] endorse) that scientists will compete for priority in solving the most significant problems of their discipline is open to question.

Table 5.5: "Selected Problem After Comparing It To Other Problems" by "Problem Suggested by Theory"

	Problem Suggested by Theory			
	Yes		No	
	Percent	Number	Percent	Number
Selected problem After Comparing it to Other Problems	67%	34	36%	19
	p =.000 (1-tail, Fisher's exact test)			
		51		53

2. Other Indicators of "Shared Problems"

My views about the meaning of the differences in proportions of selected problems compared with others before they were chosen of course are not unassailable. They represent results of data collection and analysis based on a particular way of operationally defining "shared" problems. I was not comfortable with a single indicator and decided, therefore, to try other ways of assessing the rationalist idea of "shared" problems.

In principle, it should be possible to go through current text books in various disciplines to see if they identify the problems of the discipline. However, I rejected this as unmanageable and chose instead to approach the issue from two angles: an indirect indicator of "shared problems" and a direct one. I chose "stable problem set" as an indirect indicator based on the following reasoning:

If problems are indeed *shared* as the rationalists assert, then those problems must remain unsolved for a reasonable, if not necessarily quantifiable, amount of time. Thus, a "stable set of problems" should exist for any specialty. The set is "stable" in the sense that change in its constituent elements of problems occurs fairly slowly (although what constitutes "slow change" may be a far more rapid change in some fields than in others). From my standpoint, presuming a stable set of problems is not an unreasonable presumption for the rationalists to make if one further assumption is made beforehand: problems are derived from theories. As I stated earlier (see Chapter Two), I believe that Zuckerman and other scholars in the rationalist tradition have taken this position *implicitly* (*i.e.,* they have not stated their assumption about this matter), and I have adduced some evidence in support of my opinion (see Chapter Two).

I used several indicators to assess whether respondents in my study were obtaining their problem from a "stable" set (or, more generally, if problems were "shared").

Did scientists find their problems in research publications? Did scientists get their problems from recommendations of renowned members of their discipline? Did they compare the chose problem with any others before they made their selection? Did they know well in advance of doing the research the title of the study they would be doing?[4]

To determine if scientists are working from a stable problem set of previously identified problems, respondents were asked if they knew quite far in advance what their next project would be. Therefore, I investigated the following hypothesis:

Hyp. 4.1: **Scientists choose their problems from a stable set of already identified problems.**

The specific hypotheses I tested were:

Hyp. 4.1.1a: **If the rationalist model is correct, scientists will report that they obtain the problems they will research by comparing problems with one another.**

Hyp. 4.1.1b **If the rationalist model is correct, scientists will report that they knew what their next problem would be well before they started to work on it.**

I shall only address Hyp. 4.1.1b in this section since I have already reported earlier in this chapter the findings related to Hyp. 4.1.1a (see section on theoretical derivation versus local organization of problems).

Data relevant to Hyp. 4.1.1b came from answers to question 75C of the Questionnaire (available from author) asking the respondents to provide the title and a brief description of their next project.

For this hypothesis, I had valid questionnaires and interview schedules for analysis from 62 respondents. Forty-three of them (69%) did *not* provide a title and/or a description of their next study (see Table 5.4, p.154). Did these 43 respondents not know what their next study would be about? That is possible. However, there are other possible explanations, and it is not prudent to jump to any conclusion about this matter.

One alternative explanation is that researchers are deliberately withholding information. There is some anecdotal evidence (Merton [1973; originally 1968]) providing reason to suppose that researchers might want to do this to minimize the likelihood that information of value to potential rivals will reach them in time to benefit from it.[5] Perhaps then, researcher respondents would not willingly share the title and a brief description of their project with anyone if they believed by doing so it would be given to rivals prematurely thereby giving the latter an edge in the race for priority.

However, I doubt researchers feared sharing these data either with me or my brother, (Dr. Joel Fisher), who did Canadian subject interviews. There was no reason to suppose I would reveal these data to anyone with whom my respondents were competing. Only one researcher actually showed some nervousness about my possibly betraying her confidence.

Nevertheless, to rule out any possibility that the respondents' desire for secrecy lay behind the low response rate to this item, I looked at responses of two component groups -- less productive and highly productive researchers. I reasoned that the most productive scientists were generating scientific research at a high enough rate that rivals would not gain any edge from advance knowledge of the research plans of these productive researchers.

Among the 62 respondents for whom it was possible to assess the issue the results for the more productive group were 23% not providing title and/or description compared to 38% for the less productive group.[6] The results were tested (by Fisher's Exact Test) and found not significant. As expected, therefore, it does not appear that there was any appreciable difference between them.

This crude test, (all that my data would permit), does not put the matter to rest forever. However, a more refined test will have to be deferred until a study of a larger population can be done.

Assuming, then, that secrecy does not explain why respondents failed to provide title and brief description of their next project, what other explanation is possible? I assessed two possible explanations.

One possibility is that only truly autonomous researchers might be in a position to know their next study. By this reasoning, researchers in government agencies and commercial organizations could be assumed to be non-autonomous, *i.e.*, hired hands who are given work assignments to do without any say in the matter. In other words, I could look at work setting to see if it made any difference in the proportions stating the title and/or describing their next project.

For this test, I had 58 usable cases that I divided into two groups: those with university/college affiliations and those with other affiliations.

This comparison yielded no difference in proportions providing the title or description of their next project (33% versus 32%).

I then tried a different approach. I reasoned that perhaps a more sensitive test would be a comparison between those who claimed "freedom to choose their problems" and those not making this claim. This comparison did produce a large percentage difference between those in the "free to choose" group giving a title (33%) and those in the "not free to choose" group giving a title (18%).

However, the analysis was based on too few cases for this difference to be statistically significant at the .05 criterion level (by Fisher's Exact Test).

Does freedom to choose lead to a greater likelihood of knowing one's problem in advance? This question cannot be answered with the data in hand. Yet, one should not forget that only a third of those with freedom of choice gave a title or description of their next problem. From the standpoint of the "garbage can" model, this suggests that a considerable group of researchers are not selecting from a set of shared problems known to members of the research specialty.

However, my finding has broader implications as well; *e.g.*, it suggests that Zuckerman's question as to how scientists choose from the stock of already identified problems is counterfactual for many disciplines. And by making clear that Zuckerman's visualization of the process of problem choice is erroneous to some degree, my finding casts doubt on the Kuhnian conception of science as consisting of fields characterized by dominant paradigms and a set of associated problems known in advance. Some scientific disciplines do have dominant paradigms (and these paradigms are useful sources of problems as Kuhn [1970] shows), but certainly not all do.

I must reiterate here that there is no conclusive evidence that rationalists believe that scientists have perfect information about the importance of problems beforehand and thus are able to choose the most significant to study. Yet, if rationalists are not assuming perfect information, frankly, I am perplexed as to why in her original formulation of the rationalist model Zuckerman couched her question about the process of problem choice in terms of selecting from a stock of *identified* problems. She certainly raised questions in the minds of readers about whether she was assuming scientists made choices under conditions of perfect information about problem significance. And these questions about her intent are not really dispelled by her comment elsewhere in the same article (p.80-81) that "Observations of this sort suggest that scientists focus on problems they conceive of as soluble but that they do not then and there identify all the soluble problems." Her toying with the possibility that problem choice occurred in a state of full information about the significance of problems placed her at odds with Merton's published views (1959:xi) on problem finding. "Little is known, in any systematic way," he wrote, "about the conditions and processes that lead men to find problems of consequence for science. The experience remains obscure."

Merton's words hardly suggest that the process of problem finding consists of some first rate library work, supplemented perhaps by that "*je ne sais quoi*" of scientific taste that separates the truly outstanding from the competent but undistinguished in the area of problem selection. Merton, widely regarded as one of the finest sociologists of our era, was not ruling out the need for solid library research; he was simply remarking that he believed that there were other conditions and factors involved in finding researchable questions. And he was also confessing that he and other sociologists at the twentieth century's midpoint had only a vague idea of what these might be.

I suspect that Zuckerman and Gieryn, after reading Kuhn's book *The Structure of Scientific Revolutions* (1962), concluded that most of the time deriving significant problems in the "advanced sciences" was a straight forward process of logico - deductive reasoning from the prevailing theory.[7] A good theory, one might infer from Kuhn's monograph, suggested numerous good problems which once operationalized could be tested by average practitioners of the discipline.

There were of course a few messy issues to deal with (as Kuhn was well aware), *e.g.,* the possible existence of two (or more) plausible contenders for the title of prevailing paradigm. However, Kuhn seemed to be suggesting that usually researchers in the exact sciences were working with one basic paradigm whose correctness was essentially agreed to by all.

Competition for Priority: Evidence of Interpersonal Rivalry

Besides looking at a variety of indicators of "shared problems" (*e.g.* whether the problem set was "stable"), as an indirect indicator of competition for priority, I also took a more direct approach to investigating competition for priority in science. My indicator was "interpersonal rivalry." Again, I must emphasize that I did not attempt to explore competition in depth. I wanted to see if there was evidence to support the rationalists' argument that scientists are competing for "priority" and this influences problem choices.

The reader may recall that rationalist scholars such as Gieryn (1978), who argue that problems are shared, imply that scientists will vie for priority in solving them.[4] The rationalists would agree with Gaston (1973:74 quoted in Bok [1982]) who in his study of competition among scientists compared it to "a race between runners on the same track and over the same distance at the same time." Such competition should be common in all scientific disciplines if rationalists are correct. And there should be no difference between disciplines in the percent of respondents reporting such competition. However, as I pointed out earlier, prior studies have not established that competition for priority is, in fact, common. Because of the salience of competition in the rationalist hypothesis, I preferred not to rely on only one indirect indicator of competition for priority since, despite its surface plausibility, it might have unknown problems which make it unreliable.

Therefore, I selected another indicator of competition for priority, "interpersonal rivalry." Anecdotal evidence of such rivalries is easily found in the literature. Classic examples are Watson and Crick's rivalry with Linus Pauling. (As Watson makes clear in his memoir *The Double Helix* (1968), it was more his desire to beat Pauling than vice-versa since Watson was a much younger scientist). More recently, there was the race between Craig Ventner's team of scientists and the university consortium scientists to map the gene (the race ended in a draw).

I believe that interpersonal competition or rivalry could be expected to arise if scientists were researching problems from a problem set of previously identified problems. As indicated in Chapter Two, I also believed that interpersonal competition would be most pronounced in the social sciences (and inexact sciences generally) because of the easier entry requirements to work in these areas compared to the exact sciences (*e.g.*, physics, chemistry, and engineering).

Consequently, I tested the following hypothesis:

Hyp. 4.3: **Competition between scientists in the inexact sciences will be more intense generally than in the exact sciences.**

More precisely, I tested the following hypothesis:

Hyp. 3.3.1a: **Researchers in the inexact sciences (social sciences, biology and medical sciences) are more likely to report that one of the reasons they chose their most recent problem was a desire to do the research before a rival did it.**

In addition, I tested the relationship between discipline type and desire to discredit rivals and/or support the views of friends.

Hyp. 3.3.1b: **Researchers in the inexact sciences (where problem choice is often influenced by a desire to contribute to public controversies) will report that "some scientists do studies to discredit their rivals" or "to support the position of their friends" more than researchers in the exact sciences will.**

Because I believed that differences in disciplines might reflect differences in the gender composition of scientists in those disciplines, I also tested the following two hypotheses:

Hyp. 3.4: **Women more than men will report that "researchers sometimes (or more often) do studies to embarrass their rivals" or to "support their friends' views."**

Hyp. 3.5: **Men more than women will report "a desire to do something before someone else does it was a factor in their problem choice."**

I shall address Hyp. 3.3.1a first. For this hypothesis I had useable data on 107 respondents. In order to compare disciplines I combined "biological/ medical science" and "social/ behavioral/economic" science into a category I labeled "inexact sciences" and combined mathematics, engineering, *etc.*, and chemistry, geology, *etc.*, into a category called "exact sciences."

Table 5.6 presents the findings of my analysis. First of all, just 22% (N=22) of the 104 respondents for whom I had data, stated that one of the reasons they chose their most recent problem was a desire to do the research before a rival did it.

Table 5.6: Interpersonal Competition as a Cause of Problem Selection by Scientific Discipline

Criteria	Discipline*			
	Inexact Sciences		Exact Sciences	
	Percent	Number	Percent	Number
Chose Most Recent Research Project Because of Desire to Beat Competition	14%	8	33%	15
	p =.015 (Fisher's exact test)			
	59		45	
Chose a Research Project in Past Because of Desire to Beat Competition	27%	16	51%	23
	p =.011 (Fisher's exact test)			
	59		45	

* *I also did a cross-tabulation of the same criterion variable ("a desire in part to do it before others") against the independent variable of biological sciences/medical sciences versus all others). Twenty-six percent of the biologists/medical scientists versus 44% of all others claimed to have felt this way in the past.*

Table 5.6 also shows 33% (N=15) of the respondents in the "exact sciences" versus 14% of respondents in the "inexact sciences" claimed that they were motivated by a desire to beat out a [potential] rival. (By Fisher's Exact Test these results were significant at the .015 level for a one-sided test). These results are opposite to what I had predicted.[8]

When I looked at whether in the *past* the respondents had been motivated by a desire to beat out a rival, I found similar results. Overall, 38% said that they had been motivated by this desire. In the "exact sciences" the proportion was higher - (51%) - and in the "inexact science," it was just 27% who were so motivated. (These results based on a one tail test were *highly* significant, by Fisher's Exact Test, *see* Table 5.6).

It is interesting to reflect on why there would be less interpersonal competition in the social sciences etc. than in the hard sciences. Apparently, ease of entry into a field is rather less important than whether there are problems in the field that are significant enough intellectually and/or from a financial standpoint that people will want to solve them. Also, there may be more of these in the hard sciences than in the social/behavioral sciences.

Though as I expected when comparing a lengthier past with a brief present, the proportion claiming to be motivated by competition was higher in this Table, it certainly was not at a high enough level to suggest the "intense" or "heightened" competition envisioned by Zuckerman. These findings, however, do lend some support to Merton's (1973:331) hypothesis that "kinds and degrees of competition differ . . . among specialties."[9] If Zuckerman (1988) and Bok (1982) are correct, and if as Gaston (1978) says of competition in science, it is "a race between runners on the same track and over the same distance at the same time," a race (he pointedly notes) with no awards for "place" and "show," then why was I not finding more evidence of competition?

Perhaps it is because there is less "head-to-head" or interpersonal competition than implied by Zuckerman, *etc.* It is true, of course, that there were more people doing scientific research in the United States in the late 1990's than there were forty or fifty years earlier. However, as Zeldenrust points out (1990:1) researchers are often employees of a large concern, consequently, they do not experience the competition in a direct way.

In the 1960's, Hagstrom discerned this fact as evidenced by his observation (1986:47; originally 1965) that:

> Those scientists who discover important problems upon which few others are engaged are less likely to be anticipated and more likely to be rewarded with recognition. Thus, scientists tend to disperse themselves over the range of possible problems. . . . In many disciplines, dispersion to avoid competition takes place not only over the range of problems available but over the range of institutions.[10]

Intense interpersonal competition is costly. Most researchers (in certain fields at least) are supported by government funds and to get the most value for its grants the government probably wants to avoid duplication of effort. It should perhaps also be noted that while the fact that they have competitors is not an incubus for many scientists, neither does competition motivate them. As a case in point, a prominent biologist (S070) in my study population claims that he *never* was motivated by a desire to beat out someone else. In answer to Q37 and Q38 of the questionnaire, he sneered, "It's a waste of time" either to worry about competition or let it motivate you in what you do.

However, not all those disclaiming that competition was motivational were so disdainful. For example, another biologist (S072) commented, "No. [I was not motivated in my problem choice by a desire to beat out a rival]. But [competition] became an important consideration in how much time I spent on the work. [It is] important for publication purposes and [especially to get an article into] a high quality journal for dissemination [to consider the existence of competitors]."

Some respondents, while agreeing that they were motivated by a desire for priority, did not think in terms of beating out a rival. A woman engineer (S048) said, "not really a rival but other colleagues might also be working on it," in explaining why she was motivated to work on the problem she chose. For her there was no antagonism as suggested by "rival," simply a desire to be first. And a geologist (S036) made it clear that he also had no specific competitor in mind -- "not a known rival. Simply to get there first," he explained in regard to his motivation.

Merton (1973:331-332; originally 1968) observes that except among a few elite scientists, competition in science is "more impersonal," that "scientists often do not know who else is engaged in similar work and this lack of information generates its own brand of pressures and anxieties. Competition is less often experienced as a sportive, though often intense, personal rivalry; it tends to become a diffuse pressure to publish quickly in order not to be preempted by unknown others."[11]

The economist William Baumol offers another reason why there may be little interpersonal competition among scientists. According to Michael Weinstein (*New York Times*, June 5, 1999, B7), Baumol.

> . . . shows how companies pour money not only into their own research and development but also into such operations by their rivals. Firms participate in joint ventures that hire teams of researchers to develop technologies that the firms will share. They also engage in the largely unrecognized practice by which companies enter into technology – sharing compacts.

What Baumol forcefully demonstrates is that oligopolistic competition is good for research, at least applied research, expanding opportunities for researchers to practice their craft. From my standpoint, it is difficult to imagine a researcher at company A developing any sense of personal hostility to a supposed rival firm that has invested money in research she and her colleagues are doing, simply because she knows that the rival will benefit from her work. Similarly, she and/or her firm are benefiting from research work the rival firm is doing. (If not, A's firm is unlikely to continue the technology compact since, as far as benefits are concerned, it is a "one-way street").

Competition for Priority and the "Garbage Can"

This is a good point to stop and recapitulate. Some writers on science have argued that competition in science is intense, though empirical studies I have reviewed have hardly demonstrated the correctness of this argument. Certainly scientists appreciate recognition for their efforts, and researchers only get recognition for their priority. However, the breadth of issues scientist's tackle is vast and most scientists I suspect do not experience head-to-head competition in their work. Thus, it is not a major factor for most and not at all a factor for many in problem selection.

Quite possibly, however, there are pockets of intense interpersonal competition in particular specialties. The likely candidates for finding intense competition are what Merton (1973:331; originally 1968) calls the "hot fields." Such specialties, he writes, "... tend at least for a time, to attract larger proportions of talented scientists who have an eye for the jugular, concerned to work on highly consequential problems rather than ones of less import. Hot fields also have a high rate of immigration and a low rate of emigration, again until they show signs of cooling off." Intense competition may also occur in fields that *were* very "hot" but almost overnight have "cooled off" leaving many researchers scrambling for the limited funds available. Over time, most of these scientists will migrate elsewhere, I expect, and competition among those remaining will largely dissipate.

If head-to-head competition as a source of problem choice is mostly of secondary significance, conflict at the *inter-personal* level is even less relevant. I found no evidence that conflict (between people), was a motivating factor in problem choice. One reason already was given for not expecting conflict to be of any consequence to problem choice: interpersonal conflict implies *awareness* if not acquaintance; however, the researchers do *not* know who their competitors are.

This is not always true, of course. Witness Zuckerman's comment (1988:512-513) about the practitioners of the sociology of science.

> Not unlike other specialties this one is marked also by warm and lively, if not always cordial, interaction between adherents of different theoretical orientations, no one of which holds

> sway: constructivism, discourse analysis, relativism, structural analysis, functional analysis, and conflict theory. These diverse perspectives are sometimes linked with differences in foci of attention and nationality, such that cognitive conflict is sometimes transformed into social conflict. Those focusing on the sociology of scientific knowledge - largely though far from all, being English and European researchers - have often been at odds with those focusing on the social structure of science - largely, though far from all, being American researchers.

Though I could find no marked evidence of interpersonal conflict in my study population (just one respondent - albeit a renowned mathematician[12] - even acknowledged being embroiled in a quarrel with other scientists), I was able to detect possible indications of how inter-organizational conflict might affect problem choice. Based on Hyp. 3.3.1b, I had asked a pair of questions pertaining to behaviors I thought suggested inter-personal and/or inter-organizational conflict as a source of problem choice (Q35- Q36):

A. Do some researchers choose problems for research primarily to embarrass their rivals or personal enemies?

B. Do some researchers select their problems because they want to support the intellectual positions of their friends?

In regard to "A", as indicated by Table 5.7, I found little or no evidence that scientists try to embarrass their enemies in print.[13] There is probably a norm of some sort in science that makes this highly unlikely behavior for scientists. On the other hand, many scientists indicated that supporting intellectual positions of their friends certainly was a motivating factor in choosing a problem.

Table 5.7 shows that a third (34%) of the 103 respondents for whom I had information claimed that some scientists did choose problems with the aim of supporting friends. Results may vary by discipline; 44% (N=15) cases in biology/medicine said this was true in their opinion, rather more than in many other fields. (However, the difference was only marginally significant statistically in my data set).

Table 5.7 Interpersonal Conflict as Criterion of Problem Choice by Discipline of Respondent

	Discipline					
Some Researchers Choose Problems to Embarrass Rivals or Personal Enemies	Biology/ Medicine		All Others		Total	
	Percent	Number	Percent	Number	Percent	Number
Sometimes or More Often	14%	5	19%	13	17%	18
	p = .387 n.s. (Fisher's exact test)					
	35		68		103	

Some Researchers Select Problems To Support Intellectual Positions of Friends	Biology/ Medicine		All Others		Total	
	Percent	Number	Percent	Number	Percent	Number
Sometimes or More Often	44%	15	29%	20	34%	35
	$X^2 = 2.35$ $p = .097$					
	34		69		103	

This finding about biologists takes on added significance in light of the finding reported above (see Table5.4) that in biology or medicine a large majority of researchers (66%) did *not* choose their "next" research problems based on theory commitments. Far more important for the biologists and medical researchers in their "next" problem selections were the endorsements of respected colleagues and leaders in the field (58%).

Consequently, I looked at the relationship between "supporting friends' positions" and "whether work on the problem chosen had been urged by renowned colleagues." There was indeed a positive relationship in my data between the two variables. Table 5.8 shows that 41% (N=20) of the 49 cases claiming that the problem they selected was regarded as important by renowned researchers also indicated that in their opinion researchers sometimes, or more often, did research supporting the positions of their friends.

Table 5.8: Some Researchers Choose Problems to Support Positions of Friends by Renowned Researchers Regarded Problems as Important

	Renowned Researchers Regarded Problems as Important			
Some Researchers Choose Problems to Support Positions of Friends	Yes		No/No Opinion	
	Percent	Number	Percent	Number
Sometimes or More Often	41%	20	27%	14
	p =.10 (Fisher's exact test)			
	49		52	

Twenty-seven percent (N=14) of 52 denying that the problem they chose was seen as important by renowned scientists felt the same way (see Table 5.7). (Biologists and medical researchers were no different in the proportion expressing this view than were other researchers in my study population.)

An engineering professor (S055) whose research is multi-disciplinary in nature offered some insight into this issue:

S055: There is a more subtle selection that goes on [in problem selection] having to do with 'disciplinary bias' and wanting to support the intellectual position or dominant paradigm of one's peer group.

Inter: Can you elaborate a bit on this?

S055: When I was at (a research think tank) they were interested in work utilizing (a particular discipline's theory). You just got no attention if you didn't do that. Now I am here (at his current university), and they are only interested really in complex computer modeling. Even though I think some other research (human factors in policy-making) is important there's no interest in it.

Respondent SO55, (who is probably not an unusual case), clearly is under some pressure to make use of specific paradigms, or particular techniques, whose "scientific" worth is accepted by the organization culture. He is aware that a "bottom-up" point of view might suggest that straightforward (but low technology) research would yield equally big dividends in scientific knowledge but hesitates to act on this insight.

The respondent's remarks draw attention to an important role of conflict first discerned by George Simmel. Simmel, according to Coser (1956:34), ". . . contends first of all that conflict sets boundaries between groups within a social system by strengthening group consciousness and awareness of separateness, thus establishing the identity of groups within the system."[14]

Coser elaborates on Simmel's point noting: "It seems to be generally accepted by sociologists that the distinction between 'ourselves, the we-group, or in-group,' and 'everybody else, or the others-groups, or out-groups' is established in and through conflict. This is not confined to conflict between classes."

Coser does not address organization boundary maintenance through conflict between organizations. However, it is clear that he would argue that the theory applies to organizations as well as to social classes, *etc.* He would also perhaps concur with my assertion that boundary maintenance in science organizations is accomplished by scientists doing research work no one outside of science is likely to be able to do. An example might be computer modeling - regardless of its payoff in knowledge compared to research activities less sophisticated technically, scientists will engage in computer modeling because it looks very scientific.

What are the implications of these findings on competition and conflict from the standpoint of the rationalist model and alternative models? The rationalist model's value depends to a considerable degree on its correctly predicting that there will be plenty of competition for priority in science. I believe that this should be manifested in interpersonal competition and/or "stable" problem sets. However, there is little *interpersonal* competition in science between researchers and there is no effect on problem choice according to my data. In this regard, my study's findings simply echo the case study findings in Mulkay's (1974) work (though he insists on reaching the opposite conclusion) and Zeldenrust's (1990) work in support of Hagstrom's (1965) conjecture.

Evidence for Competition for Priority

If competition for priority were common, then the following would have to be true: (1) there should be strong evidence that problems are "shared" and/or "identified" and this would suggest that all scientific disciplines have dominant paradigms. (2) there should be a feeling among scientists that there is intense competition for priority with other scientists. However, I have already presented evidence that problems can come from *anywhere*, though theoretically derived problems may be more common than those not derived from theory (e.g., prior findings *etc.*) (see Table 5.4). Furthermore, in Table 5.5, I have shown that the probability of a problem being "shared" is related to whether it is theoretically derived; those kinds of problems are more likely to be "shared" than those not theoretically derived. And now, I have demonstrated that interpersonal competition is not especially strong among scientists and that some disciplines and specialties lack a "stable" problem set of identified problems that might give rise to competition for priority.

Altogether, these findings cast strong doubt on the social constructivist position that (1) problems are locally originated and (2) that the choice process is chaotic.

The findings also have adverse implication for the rationalist hypothesis, though not the same consequences as for the social constructivist position.

To my mind, the implications of my data are that the rationalist hypothesis may be correct for an uncommon situation:

a. There is a stock of "shared" problems *and*

b. There are large rewards in prestige, fame and wealth for solving the most important of these shared problems.

The model of problem choice most reflective of what usually happens in problem choice situations is the "garbage can" model. It argues only that a choice is a linkage between a "problem" and a "demand" (and sometimes a constraint). The linkage occurs when the two streams of problems and demands (and possibly also the streams of constraints) flow close together permitting a linkage to occur.

This model allows, but does not require, that there be "multiple discoveries" (simultaneous discoveries of the same phenomenon by two or more teams of scientists working independently of each other). Therefore, it can accommodate the occasional interpersonal competition I found in my data set. The rationalist hypothesis, which I believe is correct in special circumstances, (*e.g.* "hot " fields, disciplines such as high energy physics that have dominant paradigms) predicts much more interpersonal competition than I found.

Minimal interpersonal competition does *not* imply lack of competition. There is competition of the kind familiar to economists: competition *for resources*. And one of the interesting findings of this study is that in this competition for resources, there are *patterned* differences in strategy.

In this regard, I only investigated *gender* patterned differences in problem choice criteria (see Hyp 3.4 and Hyp. 3.5 in Chapter 3).

I discuss the findings on this inquiry below where I consider gender issues in depth. Here I will simply observe that (a) there are differences in men and women respondent's views on how competition influenced problem choices (see Table 5.9). And (b) these differences should not exist if the rationalist hypothesis is correct.

Table 5.9: Interpersonal Competition as Criterion in Problem Choice by Gender

	Gender			
	Men		Women	
	Percent	Number	Percent	Number
Chose Most Recent Research Project Because of Desire to Beat Competition	29%	18	12%	5
	p = .04 (Fisher's exact test)			
		62		42
Chose a Research Project in Past Because of Desire to Beat Competition	46%	29	24%	10
	p = .021 (Fisher's exact test)			
		63		42

One final observation I want to make is that I believe that gender -based differences in problem choice criteria is *not* the only such patterned difference. I believe that there could be other such patterned differences; *e.g.,* minority scientists, especially African-American and Latino scientists, adopting strategies somewhat different from those of the white male scientists who predominate in American and Canadian science.

The socially patterned differences in problem choice strategies between genders can occur simultaneously with such other influences as the impact of the organization of orientation on problem choice (see the multi-variate *analysis infra*).

D. ***Summary***

Because Zuckerman (1978:83) had argued that scientists are drawn to solving the most important problems possible given their theoretical and methodological commitments, and, furthermore, because she had in various writings (1978;1988) spoken of the acutely competitive environment of science, it is reasonable to infer that scientists would be motivated by interpersonal rivalries in solving problems in science.

In my mind, the implication of intense interpersonal competition as scientists compete for priority in solving the important problems of their discipline is one of the most attractive features of the rationalist perspective. Therefore, finding that interpersonal rivalries are not all that common (or at least not commonly admitted) and that they do not have any major bearing on problem choice removes much of the appeal of the rationalist perspective. (Note that the desire of researchers to support the intellectual positions of friends can be interpreted as generally consistent with the researchers' expected responses [in the "garbage can" model] to their work organizations' endorsing of particular paradigms.) I shall have more to say about this point below when I address organization trait influence on research team criteria of problem choice). Taken together with the findings already reported on scientist autonomy and on where problems are found, they reinforce my conviction that a new model of problem choice is needed. The best available alternative to the rationalist model, in my opinion, is the "garbage can" model.

Yet, up to now, I have not scrutinized this model. Its virtues, other than an absence of certain defects in the rationalist and social constructivist models, have not been assessed.

The first thing to remember is that whereas the rationalist and social constructivist models see the choice of problem as an *individual scientist's decision* the "garbage can" model describes problem choice as a process at the organization level. The choice consists of a "linkage" between a "problem" and a "demand" (and sometimes a "constraint"). Scientists are conceptualized as suggesting problems in this process and looking for a "demand" in order to make the linkage. Alternatively, scientists are responding to a "demand" by suggesting a problem(s) that might satisfy the "demand." However, this process cannot be directly observed by the outside observer.

What can be observed is whether or not scientists have *internalized* the organization (of orientation's) wishes. I maintain that if organization wishes are incorporated in individual scientist criteria of problem choice, this is evidence that the scientists, as socialized members of the organization, are likely to be either:

1. detecting an organization "demand" and suggesting research problems to meet it or

2. suggesting problems and then attempt to ascertain that there is an organization "demand."

After all, unless this view is correct, why would scientists internalize organization wishes in their problem choice criteria? Certainly, there would be no need for scientists to do this according to the rationalist view. If the rationalists are correct, scientists' problem choice preferences should be explainable solely by pointing to the scientists' disciplinary background. That would be an indicator that it is the theories and methods of their discipline that totally guides their problem choice preferences. Of course, another (albeit *unlikely*) possibility is that we would not find any stable preferences among scientists. They would be opportunistically reacting to findings at their laboratories, perhaps driven by their "mania to write" (*see* Latour and Woolgar [1978]).

In the next section, however, I will look at the influence of the organization of orientation on scientists' problem choice preferences. From the data presented there, I will derive a general conclusion that the organization role in scientist problem choice preferences is decisive in many spheres. For example, this organization role includes not only blocking of certain kinds of investigations but also active promoting of other kinds of inquiries. And if this conclusion is correct, then I have taken a big step towards justifying my building an empirical theory of problem choice based on a "garbage can" model.

Organization Environment

Prefatory Remarks

Most scientists are employees, or, more generally, members of an organization. Caplow (1964:171) points out, "In order for an individual to function as part of an organization, he must accept some of its purposes as his own."[15] Put somewhat differently, organizations want their members to work towards accomplishing the organization goals. In the case of scientists, it would be reasonable to expect that the scientists would be doing research of interest to their employer. Thus, commercial organizations would want the researchers to do research that would help the organizations improve their profitability. Universities would want their faculty scientists to do research that adds knowledge of the respective disciplines of the scientists.

Organizations do not presume that their members, especially new members, understand the key goals of the organization. In order to ensure that researchers work toward accomplishing the key goals of the organization, they socialize the members. This is done in various ways: *e.g.,* (a) formal staff orientation, (b) periodic training of staff, (c) distributing policy documents to staff setting forth the organization's aims and means by which the goals can be met by its staff, including researchers. Even the equipment purchase decisions of the organizations may convey their wishes about the organization goals the researcher should work towards.

In Chapter One, I have asserted that the rationalist model and social constructivist perspectives slight or altogether ignore the crucial role of organizations in problem choice. An example of this neglect of the organization's influence on problem choice that readily comes to mind is Zuckerman's discussion of multiple discoveries in science (1988:544-545) (a symbol of simultaneous choice of the same problem). She never mentions the role of organizations in the process. The closest she comes is in commenting on "the culture base" which focuses "attention on certain problems" and "provides the necessary concepts and tools for solution."

And Karin Knorr-Cetina, one of the most perceptive social constructivists, likewise says nothing about the role of organizations in multiple discoveries. She speaks of the role of "scientific institutions" and "familiar forms of social control" which can quite possibly refer to norms of science and evaluative mechanisms in science separate and apart from research organization demands for research on particular issues.

In contrast, the "garbage can" model I advocate requires an important role for the organization in the process of problem choice. This is because problem choice is a linkage between a real demand and a problem (and sometimes also constraints). Choices occur when the independent or loosely coupled streams of problems, demands, and constraints come close together to allow linkage between them to be forged. Thus, organizations that want particular types of research performed will push the demand flow close to the problems flow for this purpose and/or will want their researchers to push the problems flow close to the organization's demand flow.

The latter possibility as a means of accomplishing linkages requires that the "garbage can" model will be concerned with the issue of how effectively organization socialization of its members is being accomplished. (After all, the organization cannot presume scientists in its employ will offer problems that meet its needs without socializing those scientists effectively.)

A rough indicator of the extent to which the organization has effectively socialized scientists is the *consistency* between various criteria of problem choice reported by principal investigators and their perceptions of organization concerns. (Note that open organizations such as universities have few demands of their own in respect to faculty research. However, organizations of orientation [sponsors paying for research such as the various federal government grant making agencies, foundations, and corporations] will have more explicit wishes about topic and means by which the research will be carried out).

The criteria of problem choice are the key constituents of the organization culture of the research team that the embedding organization can influence. This organization culture is important in my opinion to understand the control strategies that the team uses to help choose problems to offer for solving organizational needs of the embedding organization.

In this section my intent is to indicate that organizations socialize their researcher member to organization purposes in the problems the researchers suggest. In making my case for this point I will also argue that organization membership in an open organization, *i.e.,* one where the work organization and the "organization of orientation" can be differentiated for research purposes, is not an exception to the appropriateness of the Zeldenrust "garbage can" model of problem choice.

With reference to my claim about the influence of organizations in socializing scientists, I will argue that, first, the work organization loosens the researcher's commitment to the paradigm she has brought with her and socializes her to embrace other paradigms endorsed by the work organization. Second, the work organization socializes the researcher to accept other criteria of the problem's worth (such as whether patents can result from the work) besides technical criteria such as the mathematical tractability of the problem she may have brought with her. These points, I believe, will be evident from the results of my investigating the general hypothesis that certain organizational traits, possibly emblematic of the culture of the organization, predict whether certain criteria of problem choice will be used by research teams within the embedding organization. Operationally, I looked at the following specific hypotheses.

Hyp. 3.6: **There should be a significant negative relationship between the criterion "the**

problem was suggested by theory in my own field" on one hand and the organization trait of "the organization wants me to find research problems by considering whether the work has commercial application."

Hyp. 3.7: **There should be a significant positive relationship between the criterion of "I found the problem I wanted to do by reading widely outside my own field and the organization trait of distributing the organization mission statement and/or other policy statements expressly encouraging MDR by staff."**

Hyp. 3.7a: **There should be a positive relationship between the criterion of "I found the problem I wanted to do by reading widely outside my own field" and the organization trait of "the work site encourages MDR projects by (a) specifically seeking out grants for MDR or seeking funds allocated by the work site for MDR and (b) providing grants of funds specifically to help MDR projects.**

Hyp. 3.7b: **There should be a significant relationship between "perceived that the organization wants the scientist to find research problems by considering whether the problem makes use of equipment the employer wants to see utilized" on one hand and "in choosing the problem I wanted to study," I considered whether I would have to borrow research techniques from another discipline" on the other.**

Hyp. 3.7c: **There should be a significant relationship between "considered whether I could acquire special equipment I do not presently have" on one hand and "perceive**

that organization wants me to find research problems by considering whether problems require a novel analytical procedure that the employer wants to see utilized."

Hyp. 3.8: **In organizations where the organization wants principal investigators to consider the political sensitivity of the problem, the respondents are more likely to report that political sensitivity of a problem to federal, state, or local governments is important to them in problem selection than in organizations where there is no organization position on this issue.**

Hyp. 3.9: **There should be a positive relationship between the criterion "I considered whether the result of my work could be patented" and the organization trait of "the organization wants me to find problems by considering only problems with commercial applications.**

Findings

Table 5.10 shows the relationship between "organization wants the researcher to find problems by considering only problems with commercial application" on one hand and "the problem I chose was suggested by theory in my own field" on the other. There is a difference in the expected direction (13% versus 46%) in the percentage stating that "the problem was suggested by theory in my own field."

Table 5.10: Theory in Own Field Suggested Problems by Organization Wants Research of Commercial Value

	Org. Wants Researcher of Commercial Value			
	Yes		No	
	Percent	Number	Percent	Number
Theory in my Own Field Suggested Problem	13%	1	46%	36
	p = .068 (Fisher's exact test)			
	8		78	

Table 5.11 also shows that scientists working in organizations that want their researchers to consider only problems with commercial applications were overwhelmingly likely to consider whether the results of their work could be patented compared to researchers based in organizations not necessarily wanting their researchers to think about commercial applications. Fifty percent of the former versus five percent of the latter evaluated a problem for study based on expectation of whether results of the research could be patented. However, the **N** of cases is not large. Thus, though the statistical results are highly significant, it is prudent to treat these findings as only an interesting possibility (until the relationships are replicated in a study population that includes a larger group of commercial researchers.)

Table 5.11: Work Could be Patented by Organization Wants Research of Commercial Value

	Organization Wants Commercially Valuable Research			
	Yes		No	
	Percent	Number	Percent	Number
Work Could be Patented	50%	4	5%	4
	p = .002 (Fisher's exact test)			
	8		80	

The next hypothesis (Hyp. 3.7), was also examined by cross tabulation. Table 5.12 supports the hypothesis strongly. Over seventy percent (73%) of those claiming that their work site encourages multi-disciplinary research (MDR) work by disseminating policy statements to that effect versus 36% of those expressing no opinion or denying this stated that they found their latest research problem by reading widely outside their own field.

Table 5.12: Found Problem in Outside Reading by Organization Formally Supports MDR

	Organization Formally Supports MDR			
Found Problem in Outside Reading	Yes		No	
	Percent	Number	Percent	Number
Yes	73%	32	36%	16
	p = .000 (Fisher's exact test)			
	44		45	

Although it can be argued that Table 5.12 merely reflects the scientists' desires to find ways to apply their own discipline's paradigms to "virgin territory," it is more likely that it shows that these scientists are looking for new theoretical ideas in other fields to apply to their organization's problems. The finding in the next Table (Table 5.13) reinforces the possibility that scientists are looking for ideas to apply in their own specialty. I found that the organization's endorsing of certain paradigms (by making clear that it wants certain equipment utilized or certain analytical approaches used) is reflected in the criteria the researchers use to assess problems. For example, consider the criterion of "whether there is a need to borrow research techniques from another discipline" or the criterion "whether they can acquire special equipment not presently available."

Table 5.13: "Borrow Research Techniques from Outside Field" by "Organization Wants Its Equipment Used in Research"

	Organization Wants Its Equipment Utilized in Research			
	Yes		No/No Opinion	
	Percent	Number	Percent	Number
"Borrow Research Techniques from Outside Field"	69%	20	39%	23
	p = .008 (Fisher's exact test)			
		29		59

Thus, Table 5.13 shows the relationship between "organization wants problem selected based on whether the problem investigation can make use of equipment the employer wants utilized" on one hand and the "researcher's assessment of the need to borrow research techniques from another discipline" on the other. Sixty-nine percent (69%) of those stating that their organization wants problems selected on the basis of equipment utilization also stated that they considered whether they would have to borrow research techniques from another discipline to test a theory or hypothesis in their field. Only 39% of those denying (or expressing no opinion) that their organization wanted problems selected that would utilize certain equipment also considered if they would have to borrow research techniques from another discipline. (See Table 5.13)

Table 5.14 also shows the relationship between the researcher's having considered on one hand whether she could acquire special equipment she did not currently have and the *organization's* desire that "a novel analytical approach be used" in research. Sixty-one percent (N=17) of the 28 cases claiming that their organization wanted a new analytical approach used also stated that they considered whether they could acquire special equipment before choosing the problem versus 38% (N=23) of 61 denying that their organization wanted a novel analytical approach used.

Table 5.14: "Acquire Special Equipment" by "Organization Wants Research Done Using a Novel Analytical Approach"

	Organization Wants Research Using a Novel Approach			
	Yes		No/No Opinion	
	Percent	Number	Percent	Number
Acquire Special Equipment	61%	17	38%	23
	p = .04 (Fisher's exact test)			
	28		61	

Evidence that the organization socializes its scientist member by loosening their commitment to particular paradigms and research criteria is especially important as it highlights the pervasive influence of the organization. However, it is equally important to remember that this socialization does not entail complete elimination of all disciplinary influences (See Table 5.11)

The organization's wishes become integrated into the set of criteria that the researcher brings to the task of assessing problems. The socialization process results in a change in the relative importance of prior criteria the scientists uses in assessing problems, the possible elimination of some of those prior criteria and substitution of new ones.

This socialization process is neither inherently negative nor positive in its effects. However, the rationalists, who emphasized the importance of scientists autonomy in problem choice, generally saw this process in a dark light. To them, interfering in scientists autonomy was likely to lead to de-emphasizing intellectually important problems in favor of commercially profitable studies of trivial scientific importance (*e.g.*, study how to make a better tasting aspirin instead of a low cost AIDS vaccine for use in poor countries). In some cases, interfering in scientists' autonomy would lead to the wasteful search for scientifically impossible results. (A classic case was Soviet dictator Joseph Stalin's generous support for a laboratory with a special mission. Its aim was to find proof that the new Soviet attitudes Stalin was trying to foster through propaganda would create changes in people's brains that could be inherited by their progeny. Many other projects of greater intellectual promise languished for lack of funding, hurting Soviet biological research for decades.)

As the following discussion shows, however, the organization can influence the scientists to look at new issues that are more fruitful by encouraging contact with disciplinarians from other fields.

It is common in many organizations, especially those in the high technology sphere to schedule conferences involving professionals from different disciplines (*see* Lawrence and Lorsch, 1967) in order to facilitate integration. Therefore, I wondered if there was also a relationship between "organizations regularly scheduling conferences involving professionals from different disciplines" and various criteria of problem choice. Table 5.15 shows that there is such a relationship. Fifty-five percent of those stating that their organizations regularly schedule conferences between professionals of different disciplines versus 28% of those claiming otherwise (or no opinion) also stated that they chose a problem suggested by theory.

Table 5.15: Impact of Embedding Organization's Regularly Scheduling Conferences Involving Professionals from Different Disciplines On Criteria Used by Research Team to Choose Problems

	Schedules		Conferences	
	Yes		No/No Opinion	
	Percent	Number	Percent	Number
Selected problem suggested by theory	55%	33	28%	9
	p = .012 (Fisher's exact test)			
	60			32
	Yes		No/No Opinion	
Selected problem allowed me the chance to work with congenial colleagues	35%	21	55.%	18
	p = .054 (Fisher's exact test)			
	60			33

The next issue I examined was embedding organization "cosmopolitanism-localism" as an influence on problem choice criteria. The terms, "cosmopolitanism-localism" are due to Merton (1957:387-420) but, as used here, they have a meaning different from Merton's. (So influential has his formulation become that sociologists have found all sorts of novel uses for the concepts -- see Caplow [1964:196] for a brief list. In this study, I am concerned with a specific use of the terms: the organization climate with respect to encouraging (discouraging) researchers to discuss problems across agencies and/or across other disciplines. The organizations that encourage wide discussion across disciplines and other organizations are "highly cosmopolitan"; the ones not so encouraging in this regard are more "local."

I constructed a summative index of cosmopolitanism-localism based on responses to four questions, Q61A-Q61D, in the Principal Investigator's Questionnaire (*see* Appendix). While there are four logical possibilities for scores on this variable, only the categories "Very High" (Score = 4) and the collapsed category, "All Others" (1-3) were meaningful based on the distribution of cases.

Table 5.16A shows that respondents working in highly cosmopolitan organizations were more likely to indicate that "access to special equipment" was a factor in their problem choice (72%) than were those who did not work in such organizations (34%).

Table 5.16A: Influence of Organization Cosmopolitanism on Criteria Used to Choose Problem

	Cosmopolitanism			
	Low		High	
	Percent	Number	Percent	Number
Access to special equipment	34%	12	72%	49
	p = .000 (Fisher's exact test)			
	35		68	
Found problem by talking with colleagues in another field	24%	8	50%	34
	p = .009 (Fisher's exact test)			
	34		68	

There did not seem to be an obvious reason why cosmopolitanism of the larger organization has a causal influence for scientists stating that access to special equipment was a factor in their problem choice. After mulling the issues, I decided that the link between cosmopolitanism and the scientist's "considering if she would have access to special equipment" is the concept of "organizational set."[16] Caplow (1964:201) provides an excellent exposition of this important concept. "An organizational set," he says, "consists of two or more organizations of the same type each of which is continuously visible to every other. The sociology departments of major universities constitute a set . . . the members of a set are organizations, not people."

Caplow (1964:201) also points out that "some sets are much more important than others, just as some of an individual's reference groups are more important than others." Clearly, Caplow is arguing that there is more than one possible *set* for any organization and the *relative* influence of its sets can be ranked for any particular organization.

Caplow then adds an important caution. "Not every group of organizations constitutes a set [because] the organizations may be very similar in structure but blocked off from continuous communications." "Alternatively," he continues, "they may be in continuous communication but differ in some crucial way that prevents comparison between them. Comparison is the essential function of an organizational set . . ."

The concept of "organizational set" clarifies the connection between the (subjectively perceived) organization culture of the overall organization (*not* of the research team) on one hand and the criterion in problem choice of "access to special equipment," on the other. In cosmopolitan organizations comparisons between the organization and others of their kind (*e.g.,* other big government laboratories, other university chemistry departments in the set of universities of which that university is a member, *etc.*) are going on continuously, and adjustments in standing are occurring as a result. Comparisons about the equipment (faculty or employee) the scientists can access at their respective institutions are among the kinds of comparisons that managers of the member organizations in the set make.

Furthermore, once they have invested in such equipment (either to enhance their prestige within the set or keep up with the other organization members) managers want to see the equipment used.

The scientists also have a desire to take advantage of the resources offered by their organizations. First, the scientists expect that failure to use the equipment could cause administrators to be reluctant to spend money for equipment in the future since current equipment is not put to good use. Second, scientists may, within their own organizational set (research teams that are visible to one another), suffer loss of prestige for failure to utilize state-of-the-art techniques and equipment available to them. In medical research, for example, where paradigms as sources of problems are not as important as in the social/behavioral/economic sciences, equipment availability may be a *salient* factor in problem choice. A researcher may say to herself "I've got this wonderful imaging equipment here at my research center (*e.g.,* magnetic resonance imaging, electron microscopes, etc.) sitting relatively underutilized. Maybe I should think of projects that get some mileage out of it." And that could cause her to focus on the feasibility of a variety of projects that require use of these types of equipment. (Alternatively, the medical research organization executives could make the same judgment about imaging equipment under-utilization and import a scientist from elsewhere whose research requires using such equipment.)

But why would the scientist in this hypothetical example care that the equipment is underutilized? I suggest that she probably is a member of a team that is part of an organizational set striving for higher prestige (or at least to hold their own). The equipment is relevant to that striving. Or (and this is not a mutually exclusive possibility) she knows that the laboratory administrators would like to see more use made of the equipment which they acquired because they wanted either to enhance their standing in their own organizational set or at least to maintain their standing.

It should be clear by now that organizations have a sizable impact on the criteria scientists use to assess problems for study. Problem selection, in other words, is not simply problem driven. (Zuckerman was always careful not to exclude external influences in the problem choice process. However, from the attention rationalist scholars have given in their writings to (a) the characteristics of the would be problems and (b) the theories from which problems often are derived, it is easy to conclude that the rationalists did not generally believe external influences were particularly important in the process).

The importance of an organization atmosphere encouraging open communication has been long suspected by organization analysts concerned with organization effectiveness. However, no one has suggested that there should be any correlation between that organization variable and the criteria scientists use to help them select problems. However, the table below, Table 5.16B, shows that the score an organization receives on "organization encourages open communication (ORG COM)" is *directly* related to numerous criteria of problem choice of principal investigators.

Table 5.16B: Criteria of Problem Choice by Organization Encourages Open Communications

Criterion	Org. Encourages Open Communication			
	Very Much		Other than Very Much	
Talk with colleagues in own field	Percent	Number	Percent	Number
Yes	73%	41	50%	16
Total	100%	56	100%	32

	$X^2 = 4.809$ P = .028		P = .025 (Fisher's Test, One-sided)	

Talk with colleagues in another field	Percent	Number	Percent	Number
Yes	70%	39	44%	14
Total	100%	56	100%	32
	$X^2 = 5.699$ P = .017		P = .016 (Fisher's Test, One-sided)	

Study within time constraints	Percent	Number	Percent	Number
Yes	77%	43	47%	16
Total	100%	56	100%	34
	$X^2 = 8.279$ P = .004		P = .004 (Fisher's Test, One-sided)	

Available supplies & equipment for Study	Percent	Number	Percent	Number
Yes	75%	42	59%	20
Total	100%	56	100%	34
	$X^2 = 2.583$ P = .108		P = .086 (Fisher's Test, One-sided)	

Publish paper in new area	Percent	Number	Percent	Number
Yes	83%	44	64%	21
Total	100%	53	100%	33
	$X^2 = 4.14$ P = .042		P = .039 (Fisher's Test, One-sided)	

Assist in developing future research	Percent	Number	Percent	Number
Yes	90%	47	67%	22
Total	100%	52	100%	33
	$X^2 = 7.432$ P = .006		P = .008 (Fisher's Test, One-sided)	

Colleagues outside looking for help suggested problem	Percent	Number	Percent	Number
Yes	54%	30	38%	12
Total	100%	56	100%	32
	$X^2 = 2.108$ P = .147		P = .109 (Fisher's Test, One-sided)	

Problem study urged by renowned researchers	Percent	Number	Percent	Number
Yes	55%	31	34%	11
Total	100%	56	100%	32
	$X^2 = 3.593$ P = .147		P = .047 (Fisher's Test, One-sided)	

Prefer to map broad areas	Percent	Number	Percent	Number
Yes	44%	22	21%	7
Total	100%	50	100%	33
	$X^2 = 4.541$ P = .033		P = .028 (Fisher's Test, One-sided)	

"Organization of Orientation" versus Employing Organization: A Digression

I want to digress to discuss a theoretical issue that the influence of "cosmopolitanism" and "organization encourages open communications," throws into sharp relief.

A subtle issue that I have not addressed up to now is the fact that sometimes the "work organization" and the "organization of orientation" are not the same. I shall consider that here.

I have been arguing throughout this study that it is organizations that make research decisions. In the case of certain kinds of organizations, those that find research from internal resources, this is readily apparent. For instance, if a researcher in General Electric Company proposes a study, his work organization probably must decide if this study shall be done since it requires company funds to do the work.

Suppose, however, the work organization is a university whose faculty research is supported by a grant from the federal government, a private foundation, *etc.* The faculty members engaged in research are really not concerned so much with the work organization's point of view about the proposed research as they are with the potential grantor's point of view. It is the grantor organization that decides what research will be done. The grantor, in other words, is the "organization of orientation." This is true even if the university insists on certain requirements such as research on human subjects be approved by its human subjects research review board. The grantor, by making its grant conditional on certain university conditions also being met, has simply added to the conditions it sets. It remains the organization of orientation.

In theory, "organizations of orientation" in these situations can change from research project to research project. Typically, however, researchers have a fairly long term relationship with particular grantors. For example, for years the National Institute of Health funded a great deal of social science and other research. The U.S. Justice Department has funded criminal justice research for decades.

The distinction between "organization of orientation" and "work organization" is an important one in theory and in some actually encountered situations as well. However, it presents no insuperable problems for a model of problem choice built on the assumption that organizations make research decisions and researchers are participants in the process but do not control it.

The case of the "open" work organization where scientists' environment can include not only the values of the work organization but of the organizational sets of the researcher is important. The open organization, the reader will quickly see, is often not the organization making research decisions, though it may be a player in the decision being made elsewhere. For example, suppose the work organization is a university with a human subjects committee, and suppose the research choice decision is being made by a grants organization that wants the university human subjects committee to approve the research before the grants organization rules on it.

Zeldenrust does not discuss the open organization case because his archival data on Netherlands do not include open organizations in the sense I mean them.

However, as is now evident, the "open organization" case does not disprove the central point of Zeldenrust's model that the choice process is an organizational process.

Political Culture

In Chapter 2, I defined organization culture (following Sackmann [1991]) as "sets of commonly held cognitions that are held with some emotional investment" and which "are habitually used and influence perception, thinking, feeling, and action." Here I consider one subset of those "commonly held cognitions" that I label the "political culture" of the research team.

Before addressing the role of the "political culture" in particular, I want to reiterate a general point I have made already about organization culture; it *propels* the research team to seek problems of a particular type (or with specific characteristics). Gieryn's study of astronomers (1978) which described the process of problem specialty choice in that field provides some useful insights into this process of deciding which demands to meet. Gieryn shows that researchers accumulate problems only in particular specialties. Thus, they will probably only respond to demands that can be met from among those specialties in their range of interest. Or, they will probe for demand in these problem areas.

The organization culture will also help the group prioritize among the problems in the specialties of interest and affect how willing members are to expand their range of specialty interests, *etc.* In short, the organization culture, I contend, is a necessary, but not a sufficient, element in a theory of problem choice.

The "political culture" is that set of values/criteria of choice that reflects larger societal values, not simply those of the embedding organization and thus integrates the research team with the broad political/legal environment in which it is situated. For example, it might include allegiance to the idea that dogs and cats are pets, not suitable subjects for laboratory study. One of my respondents (S094) alluded to this major value taken from the larger political culture (*i.e.,* beyond the employer itself) when she observed: "you don't dare do research on dogs around here." This woman, herself, neither loves nor hates dogs. She simply was remarking that certain kinds of animal studies, no matter how justified they were on technical grounds, were not politically feasible. Animal research, at least on dogs, is a sensitive matter and generally it is not worth the trouble to pursue the planned canine research. Therefore, some other less sensitive animal, one, perhaps either (a) less plentiful (and thus more expensive to acquire) or (b) more difficult to work with in a laboratory setting, *etc.*, must be sought for the research study.

In regard to the influence of "political culture," presuming that organization culture were relevant I tested the following hypothesis:

Hyp. 3.10: **In organizations where the organization wants principal investigators to consider the political sensitivity of the problem, the respondents are more likely to report that**

the political sensitivity of a problem to federal, state, or local governments is important to them in problem selection than in organizations where there is no organization position on this issue.

To test this hypothesis I asked two questions (Q41a,b) of respondents regarding the "importance of political sensitivity to federal, state, or local governments" in their own problem selection. I had usable data from 79 respondents available to me for this analysis.

As expected, the hypothesis was supported by the data (see Table 5.15). Of the 29 respondents who claimed that their organization wanted them to consider if the issues were politically sensitive, 59% (N=17) also claimed that sensitivity of the political problem to government was relevant to their own problem choice. In contrast only 28% (N=14) of the other 50 respondents claimed that they nevertheless believed that political sensitivity of a problem to a government agency was relevant to their own problem choices.

Table 5.17A: Research Choice Criteria by Organization Wants Consideration of Political Sensitivity

	Organization Wants Consideration of Political Sensitivity			
	Yes		No/No Opinion	
	Number	Percent	Number	Percent
Political Sensitivity of Problem is Important to Research in My Field	17	59%	14	28%
	p =.007 (Fisher's exact test)			
	29		50	

Political Sensitivity of Problem is Important in Multi-Disciplinary Research	14	48%	13	26%
	p =.039 (Fisher's exact test)			
	29		50	

This finding, of course, is exactly the result predicted by resource dependence organization models (see Shenhav [1985]). It is also consistent with Fujimura (1987) and with the Lederberg-Zuckerman-

Fisher rational choice model. It is not consistent, however, with extreme forms of social constructivist theorizing, as I understand them, because it suggests prior planning in choice rather than mere opportunism in choice followed by negotiation to make the research acceptable to various audiences.

Most important, it indicates that the Zeldenrust model needs to be expanded to indicate (or make explicit) that "control" strategies are *culturally* determined by the organizational culture. (These norms should not be confused with personal preferences since, for example, the integration of the norm "no research on dogs" is not necessarily the personal preference of the researcher but a recognition that any project requiring experimental studies of dogs is a hard sell.)

In the case of multi-disciplinary research as Table 5.17A also shows, 48% (N=14) of the 29 (who believed their organization wanted them to consider a problem's political sensitivity prior to choosing it) also claimed that political considerations were important in problem choice in MDR studies. This is much more than the 26% (N=13) of the remaining 50 respondents. Although this shows the expected influence of organization climate in problem selection, it suggests that there may be a difference between within-discipline research and MDR.

In my opinion, this difference reflects either of two possibilities. One is that in multi-disciplinary research (MDR), as Shenhav suggests (1985:116), there is *external* problem control. This eliminates the need for the researcher to take into account political considerations. The only consideration for the researcher is whether he wants to participate and for many researchers, that decision boils down to a purely financial calculation. (Multi-disciplinary research participation probably *will not* contribute anything in the way of professional recognition.)[16] The other possibility is that multi-disciplinary research is not something some respondents have much experience in doing so these multi-disciplinary research inexperienced respondents who have much experience in their own disciplines are *guessing* they would not make decisions based on political considerations. However, in research in their own areas of competence they are drawing on deep experience and know that political considerations are important.

Another question (Q42) I investigated was, "What sort of impediments would cause you to want to stop, or refuse to work on a research project?" The respondents had to choose among three possibilities: (a) only technical impediments, (b) only non-technical impediments or (c) only if both technical and non-technical impediments are present.

For this analysis (See Table 5.17B), I had 77 valid

questionnaires. For statistical reasons, I combined respondents who chose either options "b" or "c" into one group. The results in Table 5.17B show that, as expected, the respondents who claimed their organization wanted them to take into account political considerations were more likely to want to stop research (or refuse to do research) for non-technical reasons or both non-technical and technical reasons (67%) (N=20) than were their counterparts in organizations where no organizational desire was found for principal investigators to select problems with political considerations in mind 38% (N=38).[17]

Table 5.17B: Researcher Reported Ever Wanting to Stop Research Project for Non-technical Reasons by Organization Wants Consideration of Political Sensitivity of Problem

	Organization Wants Political Sensitivity of Problem Considered			
	Yes		No/No Opinion	
	Number	Percent	Number	Percent
Researcher Ever Wanted to Stop Research for Non-technical Reasons	20	67%	18	38%
	p =.014 (Fisher's exact test)			
	30		47	

This concludes my presentation of evidence bearing on the role of organization characteristics in problem choice criteria.

Influence of Professional Discipline on Problem Choice

Before leaving the topic of organization influences, however, I want to briefly discuss other social influences. I have emphasized the organization's influences only because it has been sadly neglected in theorizing about the problem choice process. However, I want to acknowledge that the *discipline* of the scientists also has a role to play in the problem choice process. Particularly in the disciplines without strong paradigms (though occasionally elsewhere as well) the organization influence, I argue, *weakens* the importance of the scientist's discipline as an influence but does *not* eliminate it. A good example is the problem choice criterion of "access to special equipment."

Access to special equipment for research is an important consideration in many disciplines. When the equipment is available, scientists may well try to design studies that take advantage of the equipment's capabilities. Alternatively, its absence will force scientists to drop lines of inquiry for which that equipment is essential. A simple example might be huge telescopes such as the Keck Telescope. Astronomers offered a change to do work in the Keck Telescope might try to do studies of such phenomena as quasars, very distant objects not amenable to study without special equipment for peering deep into space. Alternatively, if they had an interest in quasars but could not gain access to the special equipment needed for their study, they might turn to other problems for which such specialized equipment is not necessary. (Many comets, for example, has been found by amateur astronomers using rather inexpensive equipment that could be bought out of their personal savings).

I explored this simple point by cross-tabulating types of discipline ("Social") and the criterion, "Access to Special Equipment Was a Consideration in Choosing My Current Project." Table 5.18 shows a strong relationship in my data set between the respondent's discipline and this particular criterion. Half (51%) of the biologists, engineers, and other natural scientists, *etc.* in my data set agreed that this was an important consideration in their research but only *four* percent of the social or behavioral/economic scientists did (Table 5.18).

Table 5.18A: Access to Special Equipment by Discipline

	Social/Behavioral/ Economic Science		All Others		
	Percent	Number	Percent	Number	Total
Access to Special Equipment	4%	1	51%	41	
Total		24		80	104
	p = .000 (2-sided, Fisher's exact test)				

Table 5.18B: Technically Superior Colleagues by Discipline

	All Others	Biologists
Low Agreement	57% (39)	32% (11)
High Agreement	43% (29)	68% (23)
Total	68	34
	$X^2 = 5.67$, $P = .017$; Fisher's Exact Test $P = .015$ (1 sided)	

Table 5.18.C: Prefer Problems No One Else is Likely to be Working On By Discipline

	Inexact Discipline	Exact Discipline
All Others	18% (10)	3% (1)
Rarely or Never	82% (46)	98% (39)
	$X^2 = 5.42$, $P = .02$; Fisher's Exact Test, $P = .018$ (1 sided)	

Tables 5.18.B and 5.18.C show that the salience of other problem choice criteria may also be influenced by the discipline of the principal investigator. This is a good point to recapitulate the main points of this section before moving to my next topic of gender.

Summary

The purpose of this lengthy presentation was to emphasize the crucial significance of the researcher's organization of orientation in the problem choice process. My main point is that the organization (of orientation) *influences the problems offered by its research scientists*. It accomplishes this outcome by first influencing which paradigm the researchers turn to as possible sources of problems (See Tables 5.10 through 5.14). And, second, the organization of orientation influences the other criteria (besides theoretical significance) that researchers apply in their choice of problems for the organization to research (see esp. Table 5.17A).

Certain characteristics of the work organization appears to be especially pertinent in accounting for (a) the particular criteria scientists apply in deciding which problems to offer for study and/or (b) explaining the manner in which they go about finding problems. The "cosmopolitanism" of the organization and/or its encouragement of communication by scientists affects certain problem choice criteria. (see Tables 5.16A and 5.16B)

Furthermore, whether the organization supports multi-disciplinary research and how it manifests that support is relevant also in explaining researcher criteria of problem choice.

On a more abstract level, the discussion thus far has bolstered the case for a "garbage can" model of problem choice by pointing to the fact that research problems which abstractly speaking, are a "solution" in March *et al.* terms are inserted at any time and/or can be withdrawn at any time and can be found anywhere (see Zeldenrust [1990], p. 22 *passim*).

The resources looking for work (real demand in my version of the "garbage can" model) are not strongly coupled to solutions (*i.e.,* problems). Because of the organization's socializing of researchers to offer relevant problems, these resources are loosely coupled to problems. The organization wants linkages to occur between demands and problems and therefore, by various means, it encourages the two loosely coupled streams to flow closely together. That is the whole point of influencing scientists to offer problems that the organization can benefit from having solved.

This concludes my discussion of the effects of organization structure on problem choice. I am now ready to present my findings on gender effects on problem choice criteria after which I will consider (in a Technical Appendix) their combined effects.

Gender

Prefatory Remarks

It may seem from the extended discussion about the impact of the organization that I am belaboring the obvious. Keep in mind, however, that there are a number of reasons for this lengthy presentation. First, the importance of organization milieu is belittled, if not ignored entirely, by the rationalist model which, as I emphasize throughout, sees problems themselves as responsible for demand to solve them. Second, the "garbage can" model gives central importance to the role of organizations in problem choice. If I am intend to demonstrate convincingly that the "garbage can" model is the appropriate basis for a theory of problem choice, I need to show that the "garbage can" model's emphasis on the central importance of the organization in the problem choice process is justified by the empirical data. Finally, a third reason is that the evidence in support of a central role for the organization in the problem choice process hints at why gender differences in problem choice and in productivity occur. At first glance, the role of gender differences in problem choice must be surprising. After all, if problem choice is a rational activity as many sociologists believe (see Gouldner[1970:55]) why would men and women be selecting different problems for research?

Rationalists have had a ready answer for this question; women are choosing different problems from men because women scholars are less free to choose, or have less access to resources than men. Unfetter women scholars and the gender differences will disappear, according to the rationalists.

In contrast to the rationalist hypothesis, the "garbage can" model makes no prediction whatsoever. It neither expects gender differences to affect problem choice nor rules out that possibility. However, in its emphasis on the nonrandom, *non rational*, and *cultural* aspects of the problem choice process, it allows for a result that would seem inconceivable to the rationalists. This result is that both men and women relatively free to select problems would select problems more dissimilar from one another's choices than those selected by men and women who are less free to choose.

There are some studies in the literature supporting the rationalist point of view that gender differences only arise from differential constraints and would not occur if scientists regardless of gender were free to choose. Feldt's study (1986) and an MIT Report (1999) both show women given fewer research resources than men. In my own study, I found that women in my study population reported even less freedom to choose their problems than men respondents. (See Table 5.19) Thus 39% of the women versus 16% of the men scientists claimed that they had to alter or forego doing a study at some point in their careers because of institutional controls.

However, I also found that when perceived autonomy was controlled, women differed from men more in their problem choice criteria when free to choose than they did when less free (See Table 5.3A and 5.3B). This is the exact opposite of what rationalists might predict.

It is unwise to build a castle on this one finding however interesting it may be. More research is needed to corroborate the finding. It is remarkable however, that no other research along these lines has been done. I think this is because the rationalists were so convinced that differences in problem choice were merely results of constraints on the freedom of woman scholars that they did not investigate these gender differences intensively. Therefore, they failed to discover that the differences are greatest among the most autonomous scientists and, of course, they did not consider the implications of this fact for the structure of science -- a striking example of Zuckerman's point that a way of seeing also is a way of not seeing (Zuckerman 1978).

Before I turn to my findings, I need to remind the reader that gender, as stated earlier, is not to be confused with sex. Sex differences are exclusively biological in nature *e.g.*, the result of purely biological processes. Gender differences *include* cultural and socialization differences of males and females. Giele (1988:309) says that "most social scientists take the view that the study of individual sex differences to be adequate, must consider both culture and innate disposition."

Gender differences, I found, sometimes have an independent effect on problem choice criteria and sometimes have an effect through their interactions with other structural features of the world of scientific research, *e.g.*, characteristics of the organization in which research occurs.

I do not probe the basis for the gender differences in problem choice that I found in this study. However, as I will note below, I share some views of Cole and Singer [1991]. I also differ with them to some degree as I make clear in my discussion on the gender productivity gap in science (see *infra*). (These differences, discussed in Chapter Five, relate to the role of variables other than socialization in the productivity of scientists).[18]

Competition and Gender

I first examined whether women were more (or less) motivated by competition than were men. As I pointed out earlier, competition is of keen interest to rationalist scholars of science, a subject to which they have given considerable attention and about which they have a distinctive position.

In my inquiry into competition, I wanted to know, among other things, (a) if interpersonal rivalries were important to scientists generally in problem choice and (b) if men and women differed in the degree to which competition influenced their problem choices. Since I earlier offered possible arguments why one or the other gender might be more interested in interpersonal rivalries, I chose a two-tail test for my hypothesis.

Table 5.9 (*supra*) shows that men were somewhat more likely (29%) than women (12%) to say that in their most recent research they were motivated by a desire to beat out a rival (albeit not necessarily a *known* rival, as indicated earlier). Still neither gender is highly motivated by competition with others, known or not. The large majority (three-quarters of the total) assert that they did not take the possibility of competition into account.

Even when I asked about competition in the past, a sizeable majority of both genders did not take the possibility of competition into account although nearly half of the males (46%) versus 24% of the female scientists claimed that at some point in the past they took competitors into account.[19]

These results, I assert, show a gender difference that is interesting in its own right. (Though it may be possible to offer an alternative explanation for each table.) However, they may also have significance for Zuckerman's views about competition in science. Recall that she (Zuckerman, 1978) saw competition inevitably arising among scientists as they scrambled to solve the most important problems before others did. And Zuckerman (1988: 538) also distinguished two classes of scientists: (a) those willing to (or keen to) engage in competition for priority in solving the major problems and (b) those shying away from the rigors of competition, who were content to solve problems of lesser significance.

It is possible to see the tabular results as indicating that the scientists for whom competition was important in problem choice are those who are eager to solve the major problems. The others, according to this line of reasoning, are content to solve problems of lesser importance. However, I think that to reason this way would be a misreading of the meaning of the results.

What I believe Table 5.9 is actually saying is that interpersonal rivalry is not a major influence on problem choice for most scientists. However, there is a ubiquitous awareness among research scientists that they are competing for scarce resources.

Gender Based Problem Choice Strategies of Scientists

While all scientists are aware to some degree that they are in a competition for scarce resources it does not follow that they utilize the same strategies for gathering the necessary resources for their scientific work. I wanted to know if there were patterned differences in research strategy. Prior research (*e.g.,* Valian [1998]) showed that there were differences in productivity between men and women but no one to my knowledge had even speculated that there might be differences in research strategy between men and women. And, of course, no one had advanced the idea that differences in research strategy might be correlated with differences in productivity. Finding such a correlation, of course, would be an intriguing clue that perhaps gender differences in research strategy help account for widely reported findings of differences in productivity. (Of course, it is a long way from demonstrating a correlation to showing causality *and* determining the direction of that causality. Indeed, if there were a correlation, it could plausibly be argued that productivity differences might cause differences in research strategy.)

There were hints in the literature that women could be expected to differ in their research planning strategies from men. For instance, in his magisterial work on the influence of birth order, Sulloway (1996:5) argued that the second born needed to choose a different strategy than the first born to get investment of parental resources in her. This could analogously apply to the situation of women in the professions. As later arrivals than men in research professions, women might need to adopt different strategies to get investment by funding sources in their work.

I investigated the following hypothesis:

Hyp. 3.11: **Men more than women are likely to report access to special equipment was a factor in their problem choice.**

Table 5.19 shows that men (50%) far more than women scientists (25%) took into account whether they would have access to special equipment in deciding if they would choose a problem to research.

Table 5.19: Criteria of Problem Choice by Gender of Scientists

	Gender			
	Men		Women	
	Percent	Number	Percent	Number
(a) In choosing the problem, I considered whether I would have access to special equipment that is not always available to me	50%	32	25%	10
	p = .009 (Fisher's exact test)			
		64		40
(b) For your next project, will this criteria be important in choosing the problem; respected colleagues in my field urged work on the problem	34%	19	68%	25
	p = .001 (Fisher's exact test)			
		56		37

(c) I chose the problem I decided to work on because it gave me a chance to work with experts in the field from whom I could learn a lot	52%	33	51%	21
	n.s.			
		63		41
(d) I prefer problems that are related to public concerns or controversies	52%	34	79%	30
	p = .006 (Fisher's exact test)			
		65		38
(e) Novel Analytical Approach to Data	59%	37	37%	15
	p = .022 (Fisher's exact test)			
		63		41
(f) In choosing problems, I wanted to contribute to a publication in a new dynamic area	67%	42	46%	18
	p = .033 (Fisher's exact test)			
		63		39
(g) In choosing problems, I wanted to contribute to something that would develop future research	73%	45	54%	21
	p = .042 (Fisher's exact test)			
		62		39
(h) In choosing my most recent problem for research, I considered whether the problem could be studied within time constraints that are applicable	54%	35	81%	33
	p < .000 (Fisher's exact test)			
		65		41
(i) In choosing my most recent problem, I considered whether adequate technical skills and knowledge would be available	85%	55	98%	40
	p < .025 (1 tail) (Fisher's exact test)			
		65		41

These gender differences were *not* affected by respondent discipline as shown in Table 5.19A and 5.19B. A possible explanation is that such gender differences could be the result of differential socialization of men and women (or even of genetic differences).

Table 5.19A: Access to Special Equipment by Gender: Inexact Science

	Men	Women	Total
Yes	48%	28%	
	(14)	(8)	22
	p = .088 (Fisher's Exact Test) (one-sided)		
Subtotal	29	29	58

Table 5.19B: Access to Special Equipment by Gender: Exact Sciences

	Men	Women	Total
Yes	51%	18%	
	(18)	(2)	20
	p = .053 (Fisher's Exact Test) (one-sided)		
Subtotal	35	11	46
TOTAL	64	40	104

This gender difference could be a plausible factor in the gender gap in productivity. If it could be shown that this is one of the socialization differences that are influential in gender differences in productivity of men and women scientists, this would go some distance in explaining why Cole and Singer's (1991) hypothesis of "limited differences" should be correct.

I also looked at other desiderata of problem selection to see if there were gender differences. Table5.19 shows there appeared to be several such gender differences.

Thus, Table 5.19 shows that:

1. Women (68%) reported more than men (34%) that for their next project, it will be important to me that respected colleagues urged work on the problem (Table 5.19(b)).
2. Women (79%) reported more than men (52%) that they usually or more often, preferred to do studies related to public controversies (Table 5.19(d)).
3. Women (25%) reported less than men (50%) that in choosing the problem they were currently studying "I

considered whether I would have access to special equipment that is not always available too me" (Table 5.19(a)).

4. Women (81%) more than men (54%) stated that in choosing my most recent problem for research "I considered whether the problem could be studied within applicable time constraints" (Table 5.19(h)).

These results taken together suggest that women are pursuing a different strategy for obtaining investment in their research work than men are using. This female strategy may reflect the problems women have experienced more than men, *e.g.*, organizational interference in their research (Table 5.20) and greater resource deficits (see Feldt [1986] and MIT Report [1999]). And perhaps it also reflects a belief (grounded in fact in some cases) that their research will be scrutinized more carefully than men's to see if it is truly "scientific" work.

Table 5.20: Ever Had To Alter or Forego Doing Study Because of Institutional Controls by Gender

	Men		Women	
	Number	Percent	Number	Percent
Ever had to alter or forego doing a study because of institutional controls	9	16%	14	39%
	p =.014 (Fisher's exact test)			
	N = 94			

In conclusion, I would say that Zuckerman's interest in classifying scientists by their attitude toward competition is misdirected. A more important point in understanding the structure of science is the gender differences in scientist strategy for acquiring research resources.

Participation of Women and Men in Multi-disciplinary Research

I next examined the relationship between gender and participation in multi-disciplinary research (MDR). Since I assert that women pursue different strategies to get investment in their research, I wondered if this would be manifested in attitude towards participation in MDR. I looked at two specific relationships:

- Belief that participating in MDR would advance my career

- Willingness to work on MDR even if not paid for it

Both the first and the second attitude were measured as a seven interval scale with an extra score for non -response. Consequently, the highest scale value was eight whereas the lowest for non-response was one.

Table 5.21 shows that women did not expect to benefit from MDR in proportions larger than those of men. (t= -1.191, df= 97 p= .236). However, I also found that women scientists were somewhat more likely to be willing to do MDR than men scientists for no compensation. This is contrary to what I expected. These latter findings were based on far fewer cases (N=82) than the former findings, are only marginally significant statistically, and may have been affected by sample attrition. I am therefore inclined not to see the results as consequential until they are corroborated in a future study.

Table 5.21A: Attitudes Toward MDR by Gender of Scientists

	Gender	
	Women	Men
Major Role in MDR - I will advance professionally	X = 4.40 (N = 60)	X = 3.87 (N = 39)
	t = 1.191 (df = 97, p = .236 (two-tail)	

Table 5.21B: Attitudes Toward MDR by Gender of Scientists

	Gender	
	Women	Men
I would do MDR for no pay if necessary	X = 5.56 (N = 25)	X = 4.91 (N = 57)
	t = 1.967 (df = 80, p = .053 (two-tail)	

This concludes my presentation of findings from the survey pertaining to the relationship between gender and problem choice variables. What can be said at this point?

There is *suggestive* evidence that indeed gender is related to problem choice variables. And the evidence has an air of plausibility, *i.e.,* face validity. However, that is all that can be said at this point.

Can we be certain that even in this small data set gender is a relevant variable in the problem choice process? Indeed, can we have any confidence that there are variables explaining aspects of the problem choice process *other* than the theories, paradigms, conceptual schemas identified by the "rationalist" camp scholars? I believe the answer to that question is a qualified affirmative. By means of logistic multiple regression analysis, it was possible to assess in a limited, but adequate fashion, whether certain non-rational variables do play a role in the problem choice process. However, the small **N** limits what it was possible to do and any inferences that can be drawn.

Multivariate Analysis

Experts in regression analysis (*e.g.,* Achen [1982:51-52]} emphasize that model building begins with a hypothesis that says which variables are likely to be important for theoretical reasons. Otherwise, model building is likely to lead to scientifically absurd results that nonetheless are mathematically satisfactory for the data set being examined. I have already indicated that I was concerned with the role of external factors, including gender and organizational variables, in problem choice since there were hints in the prior literature of their importance. (On the role of external factors, generally, see Zuckerman [1978]. Historians of science have also illuminated the role of external factors in problem choice. See, especially, Shapin [1971; 1972], Westfall [1977], and McLeod [1977]. Also see Shenhav (Ph.D. diss. 1985). On the government's role see Michael Useem [p1976b], also Deborah Weinstein for "administrative role" [1978]. Gender's influence in problem choice is suggested by Feldt [1986], Cole and Singer [1991], M. F. Fox [1992; 1995] and others. Disciplinary characteristics as predictors of problem choice characteristics is suggested by Latour and Woolgar [1979] among others).

Before I turn to the description of the variables used in building the models of problem choice criteria of principle investigators, I think a few words are in order about what I see as the relationship of these results and the earlier extensive presentation of cross tabulations. The cross tabulations presented variables that I believe could be predictors in a study with a sufficiently large sample of cases. The critical point is that the present study, based on a fairly small number of cases, does not demonstrate the importance of these predictors. They are overshadowed in the models that my colleague and I built by other stronger predictors. My inclination is to see the resulting models as demonstrating that a final model would probably contain these predictors but possibly others as well. This, of course, is a testable hypothesis.

The conclusion that I hope the reader will draw is that no conflict exists between what I have said up to now in my findings section and what I report in this regression analysis. I am optimistic that the variables in the "best" multivariate models will probably be found to be powerful predictors in a larger study replicating the present one. The variables used in the univariate analyses above may also ultimately be found to be important to such a large study. However, they did not meet the criteria for being in the best models in this pilot regression study. In any case, my main point is that gender and organization structure are variables relevant in these problem choice criteria. That will be clear from the results of the multivariate analysis as well as from the univariate analyses.

Building the Model

A. The General Conceptual Schema

I began with a conceptual schema that linked (1) organizational characteristics to particular perceived organizational preferences and to scientist preferences and (2) scientist demographic characteristics to perceived organizational preferences and to scientist preferences (see chart below)

Table 5.22: Conceptual Schema for Multivariate Analysis of Problem Choice Process

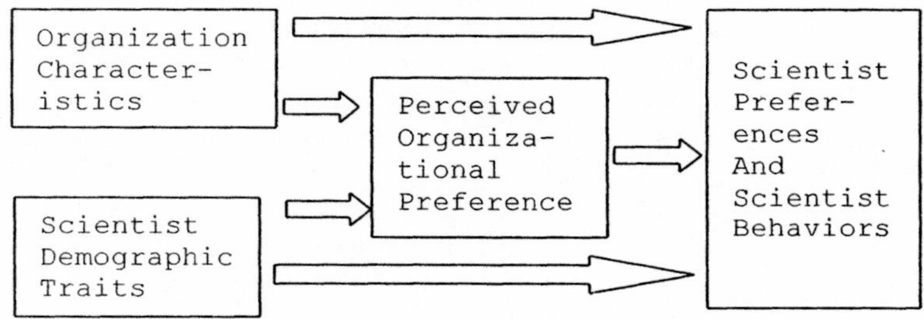

B. Filling in the Boxes of the Conceptual Schema

The organizational characteristics that I believed were relevant included:

(1) organizational support for multidisciplinary research
(2) organizational encouragement of communication among scientists.

Consequently, I selected the following predictors measuring organizational climate for examination:

a. "org. encourages multi-disciplinary research (MDR) by arranging offices for chance meetings."
b. "org. encourages MDR by scheduling conferences."
c. "org. has an organizational mission statement encouraging MDR by its scientists."
d. "org. accepts the reality of multi-disciplinary research (MDR)."
e. "org. wants scientists to find suitable research problems by talking with same discipline colleagues within the organization."
f. "org. wants scientists to find suitable research problems by talking with colleagues from a different discipline within the organization."
g. "org. wants scientists to find suitable research problems by talking with same discipline colleagues from outside the organization."
h. "org. wants scientists to find suitable research problems by talking with colleagues from a different discipline from outside the organization."

The first four variables measure aspects of the organization's commitment to supporting MDR by its scientists. The next four measure the organization commitment to communication among scientists with other scientists within or without the organization.

I also selected some computed variables based on the latter four:

1. COSMO A2 (an index variable)
2. Factor 1 (a result of principal components analysis of the four variables)

I also initially considered but rejected the following preferences:

- Prefer to do what others in field are doing
- Prefer problems no one else is likely to be working on

- Desire to contribute something of national value
- Desire to publish in a new dynamic area

after some preliminary assessment of them as problem choice predictors. (I did use them in testing models of productivity. See Chapter Six.)

The scientist demographic characteristics I believed relevant were:

a) Gender
b) Disciplinary Type
c) Years of Experience

I also selected the following dependent variables:

- Could work with experts in the field from whom I could learn;
- Could associate with technically superior colleagues (in MDR);
- Could study the problems within applicable time constraints;
- Special equipment and/or supplies was an important factor in choosing the problem;
- Considered whether I would have to borrow research techniques from another discipline to test a theory or hypothesis in my own field;
- In choosing the problem I considered whether I would have access to special equipment;
- The problem was suggested by theory in my own field and
- Chose problem after comparing it to another.

For reasons of economy, I dropped the first two dependent variables from this analysis: "could work with experts" through "could associate with technically superior colleagues" (although I used them in the productivity analysis).

C. The Strategy of Analysis[20]

The intent of the analysis was to assess the association between each outcome (*i.e.,* dependent variable) and various *subsets* of the predictors. This strategy resulted from the inherent limitations of the small data set: in every analysis done there were under 100 observations and twelve predictors plus a constant term (=thirteen parameters).

The four variables "org. wants problems found by talking with colleagues" originally had three values: 1=yes, 2=no and 3=don't know. These were recoded as 1=yes and 0=other. Thus, one must now interpret 1=yes and 0=not yes. This recoding was necessary because the number of parameters had to be reduced to obtain convergence for the maximum likelihood estimates. SEX was also recoded as 1=male and 0=female. The four variables "org. encourages MDR projects by arranging offices" or by "regularly scheduling conferences" etc. referred to above were recoded into a 1, 2 format.

Since all dependent variables are dichotomous and all of the predictors are dichotomous, logistic regression was an appropriate model. A standard statistical software package (SAS Proc Logistic) was used for the modeling effort. When the eleven predictors were used, the convergence criterion for the maximum likelihood estimates was usually obtained. When a larger predictor set including the specified eleven were used for a model, typically the convergence criterion was not met. EXP2 and EXP3, the discretized YEARSEXP was used because it reduced the number of parameters and reduced the influcence of some of the extreme values for the number of years of experience. The predictor set was: (1) org. arranges offices, (2) org. schedules conferences, (3) org. has mission statement, (4) org. accepts reality of MDR, (5) the four variable set of "org. encourages problem finding by talking to same discipline colleagues in office through org. encourages problem finding by talking to different discipline colleagues in different offices," (6) SEX, (7) a variable describing the type of scientific discipline ("EXACT") and various versions of the years of experience ("EXP2, EXP3"). Other predictors were redundant.

From a statistical standpoint, while it was often possible to obtain a maximum likelihood estimate for the model containing all variables, the covariate patters were usually too small : np<5. Thus, while parameter estimates were possible for the full model, asymptotic tests and confidence intervals were not reliable. Software that performs conditional logistic regression was examined and the problem was very large for conventional computing resources. We solved this problem in these analyses with the all subsets regression option in SAS's *Proc Logistic*.

The all subsets regression option examines for each k=1,2,3,...,12 all possible models having k predictors: performs a score test that all parameters are zero: and uses the values of the score statistic, computed under the assumption that all coefficients are zero is rejected when the score statistic is too large when compared to a chi-square distribution with the appropriate degrees of freedom. The output for this procedure lists models having k predictors in descending value of the score statistic. The better models have larger values for the score statistic.

In the analyses performed here, the same variables appeared in the best k=1, k=2, k=3, and k=4 lists of models. Typically an analysis using the best k=4 model was done followed by a likelihood ratio test for each of the four variables to assess its statistical significance and relative importance. Variables not significant at the .10 level were removed. This process continued until there remained one or two significant predictors, and possibly one other variable identified as a marginal predictor. The advantage of this strategy was that only four variables were considered at a time and tests performed on models having three variables: typically eight covariate patterns. Thus the potential for inaccurate asymptotic tests is greatly reduced and eventually eliminated with models having two variables. The model presented in each analysis is typically the best k=1 or k=2 model. By examining the output for the all subsets regression, called the "score procedure" by SAS, on can find other models and other variables that create models that are nearly as good as the one presented*. In particular there are usually several models with k=2 that are nearly as good as the one selected and they indicate the importance of other predictors. Forward stepwise and forward/backward stepwise procedures confirmed the selection of the final model.

The deviance and Pearson tests for goodness of fit were performed for each model and the covariate patterns were examined for extreme values. The final model for each analysis was found to be predictive of the specific outcome by the likelihood ratio and score tests.

D. Analysis of the Dependent Variables

For Next Project: Special Equipment is important (referred to here as "V76A")

The final model had one predictor: "Talk with colleagues outside discipline but within organization. However, SEX was a marginal predictor: ln (odds for V76A-1)=-.6931 + 1.3564(Talk with colleagues outside ….. Here ln(odds)=1n(p/1-p) and [=prob (V76A-1|V61A). Thus, there is a 98% increase in the likelihood that a respondent will consider special equipment or supplies important in the next project when they believe that there organization encourages talking with colleagues outside

---

* Interested readers can write to the senior author (Robert L Fisher) through the publisher or to Dr. Lawrence Lessner to obtain the details.

their organization. The model did not seem to be very predictive of V76A: 37.3% concordant and 9.6% discordant.

For Most Recent Project Access to Special Equipment Important (referred to here as "V21m")

The final model contained "Organizational Mission Statement Encourages MDR" and SEX: 1n (odds for V21m-1)—1.3276 + 1.2266("Most Recent Project Access to Special Equipment Important") + 1.1011(SEX). Review of the prediction patterns for each observation showed that 58.4% were concordant and 19.1% discordant. Being male and having an organizational mission statement that encourages MDR promotes the likelihood that the respondent consider access to special equipment. Four comparison conclusions follow. Among respondents whose mission statement did not expressly encourage MDR, male respondents were 112% more likely than female respondents from organizations whose mission statement did expressly encourage MDR, male respondents were 54% more likely than female respondents to consider access to special equipment for the MDR. Among male respondents, there was a 65% increase in the prevalence of men who considered access to special equipment when their organization's mission statement expressly encouraged MDR over men whose mission statement did not expressly encourage MDR. Among female respondents, there was a 127% increase in the prevalence of women who considered access to special equipment when their organization's mission statement expressly encouraged MDR, over women whose mission statement did not expressly encourage MDR. There was a 249% increase in the likelihood that a respondent would consider access to special equipment important due to being male and working for an organization whose mission statement expressly encouraged MDR.

For Most Recent Project Problem's Attraction was that it permitted use of special equipment (referred to here as "V22c")

The dependent variable V22c had 23 missing values, and there were 60 observations in the final data set. The final model contained "Org. has a mission statement encouraging MDR" and "Org encourages talking with same discipline colleagues within organization." Review of the prediction pattern for each observation showed it had 59.9% concordant and 14.5% discordant: In(odds(V22cbc2=1))=-.9404 + 1.4796(Org. has mission statement encouraging MDR) - .7008 (Org. encourages talking with same discipline colleagues...). A "yes" on V59c was that a respondent had a work site with an organizational mission statement that specifically encouraged MDR; the meaning of (Org. encourages talking to same discipline colleagues within org.)=1 is yes to "does your organization encourage talking about issues with respected colleagues in your field within the organization. From the model V59c promotes a respondent to reply yes to the selection of a problem that permits use of special equipment and, strangely, "Org. encourages talking to same discipline colleagues within the org. *reduces* the likelihood of a yes on V22c. The fact that a respondent is less likely to consider such a problem when their work site encourages interaction with respected colleagues is puzzling. Possibly it means that respondents expect respected colleagues *within the organization* to be resistant to research that is innovative in its equipment usage. But for why this should be I have no ready answer. The frequency distribution for the four variable set of "Org. encourages talking with colleagues within discipline within org. through outside the discipline outside the org., conditional on answering yes to Org. encourages talking with same discipline colleagues within org. shows that nearly all such respondents also responded yes to the other three stimuli.

There was a 125% increase in the prevalence of respondents who choose problems requiring use of special equipment problems when they came from an organization whose mission statement encouraged MDR. There was a 52% decline in the prevalence of respondents that chose a politically sensitive problem when they came from a work environment that allowed talking to colleagues in their organization. For respondents whose work place allowed talking to colleagues on site, there was a 183% increase in the prevalence of persons selecting a problem requiring use of special equipment resulting from belonging to an organization whose mission statement expressly encouraged MDR.

For most recent project considered whether I had to borrow research techniques from another discipline (here called "V21L")

The dependent variable V21L, whether the respondent had to consider borrowing research techniques from another discipline to test a theory in their own field was modeled by the predictor (org. mission statement explicitly encouraging MDR). The final model was ln(odds(V21L=1)).7307 + 1.3368*V59c, which had 43.5% concordant and 11.4% discordant. No other predictor seemed to be as useful as V59C: see the K=1 table from the selection=score output. Clearly having a mission statement that expressly encourages MDR promotes respondents need to consider borrowing techniques for their MDR when they came from an organization whose mission statement encouraged MDR.

For the most recent project ability to study problem within time constraints was important criterion (referred to here as V19time)

The dependent variable, V19time, was the statement that the problem should be studied with in the time constraints. The final model contained "org. encourages talking with colleagues from other disciplines who are outside the org" and SEX: 1n(odds(V19time=1))=.3882 + 1.6011*V61D – 1.2801*SEX, which had 62.7% concordant and 15.2% discordant. Thus working for an organization that allows discussing research with colleagues from other disciplines and outside the organizations (yes on "org encourages talking with colleagues...") and being female promote the response that the problem should be studied with in the time constraints. The predictor EXACT was a marginally significant predictor: p-value=.13. So EXACT was not included in the model. SEX was also a marginally significant predictor, but slightly more significant: p_value=.10.

For women, being allowed to talk to colleagues outside the organization increases the prevalence of the importance of time constraints by 47.7% more than for men being allowed to talk to outside colleagues from other disciplines and/or outside the organization; among respondents at organizations which do not want their employees finding projects by talking to colleagues outside the organization and/or outside the discipline, women are 105% more likely to be concerned with time constraints than men. For respondents who work for organizations that do allow them to discuss issues with colleagues from other disciplines and/or outside the organization, women were 31% more likely to be concerned with applicable time constraints then men at similar organizations.

For the most recent project, one of the criteria for choosing it was that the problem was suggested by theory (referred to here as V21G)

The outcome variable V21G was the statement; the problem I choose was suggested by theory. The final model contained predictors (org. has mission statement encouraging MDR) and SEX, and had a 57.6% concordance and 19.6% discordance. The final model was the best model with two predictors; the next best model, (org. has mission statement ...) and EXACT, was rated substantially lower then the final model. The final model was 1n(o0dds(V21G=1))= -1.2746 + 1.1955(org.

has mission statement) + 1.0610(SEX). Thus being male and working for an organization whose mission statement expressly encourages MDR promotes the selection of a problem suggested by theory by the respondents.

For women, working for an organization whose mission statement encourages MDR increases the prevalence of choosing a problem suggested by a theory by 120% over women whose organization mission statement does not explicitly encourage MDR. For men, belonging to an organization whose mission statement encourages MDR increase the likelihood of selecting a problem suggested by theory by 63% over men whose organizational mission statement does not expressly encourage MDR. Interestingly, among respondents whose organizational mission statement does *not* encourage MDR males were 104% more likely to choose a problem suggested by a theory than women in similar organizations. I do not have sufficient data to explore this in depth but it is possible that the men in such organizations are in disciplines where theory is usually the source of problems. On the other hand the women in such organizations not encouraging discussion with outside organization disciplinarians may come from disciplines where theory is weak and must be content to do studies without much theoretical significance.

E. Conclusions of the Multiple Regression Analysis

The data strongly suggests that problem choices cannot be understood without reference to organizational variables. However, gender is also clearly necessary to understanding the process. Women have evolved strategies that differ from those of men scientists in order to give themselves some advantages in the competition for resources that inevitably occurs in science. Men scientists more than women generally favor studies using sophisticated equipment even when controlling for discipline. Women rely on meticulous planning aiming for theoretical and methodological rigor even though this may be time consuming and adversely affect their productivity. Since women appreciate the time investment they must make to do high quality studies they seem more sensitive to tight time constraints than men do. Evidence from both the cross-tabulation analysis and the multi-variate analysis strongly argues that, far from eliminating differences in problem choice criteria (and probably therefore in choices) greater professional autonomy for both men and women scientists may accentuate the differences in their problem choice criteria.

SUMMARY AND CONCLUSIONS

In this concluding section of the chapter I shall review the main findings with an eye towards their implications for settling the question of which theoretical model of problem choice is superior and secondly for explaining why gender based differences in problem choice occur. Before I turn to the review, however, I want to restate concisely some principal points about the various models ("garbage can" model, *etc.*). I shall begin with the rationalist model.

The rationalist model, the leading current hypothesis, assumes that scientists want to solve the most important problems that they are able to. Thus, given freedom to do so, they try to work on the significant problems of their discipline, enduring competition for priority with their colleagues also attempting to solve those same problems. For the most part, the problems scientists try to solve are theory derived problems associated with the main paradigms of their discipline. Only rarely do scientists inductively originate a problem according to the rationalist perspective.

The main current rival to the rationalist model is the social constructivist perspective--not a coherent, unified model though scholars working in this tradition share some views. For example, one view common to all the social constructivists is their rejection of the claim that there are core ideas in a discipline. The social constructivists prefer to see all scientific facts as socially negotiated. As Knorr-Cetina [1983:169, quoted in Zuckerman, 1988:555] puts it, "... the 'cognitive' core of scientific work appears to be thoroughly social." The social constructivists also do not assume that scientists are autonomous; indeed, they argue that demand from outside science is a relevant factor in problem choice. However, they see all problem choices as locally originated--the unique result of the interaction of scientists, machines, and laboratory environments.

The third perspective on problem choice is the "garbage can" model which is heavily influenced by theoretical work of March and his collaborators on decision making in organizations. In contrast to the rationalist model, the "garbage can" model makes no assumption that scientists know what the significant problems are. The model follows March, *et al.* (see Cohen, March *et al.* 1972; March and Olsen, 1976) in emphasizing that ambiguity surrounds decision making. Ambiguity is much more than mere uncertainty about the value of payoffs from a particular decision. Ambiguity implies that many aspects of the decision are uncertain or unclear. For example, while people commonly think of solutions as following from problems, that is not necessarily true.

Because ambiguity exists, a solution may be offered even before a problem is known to exist; in effect, the solution may need to seek a problem. But that is not all. Resources may seek work. Furthermore, these processes occur without any guarantee of success, *i.e.,* resources may not find work and solutions may not find problems. This is because the elements such as solutions, resources, problems, inserted or withdrawn in no particular order, can "come, go, and wait." Hence March, *et al.*, speak of the "uncoupling" of resources *etc.* A decision occurs in these conditions when these uncoupled flows come close together and a linkage occurs between the various flows.

Zeldenrust's model of problem choice in science is March *et al.* "garbage can" model applied in a new context. If March's term "solution" is understood properly, it means anything that meets a presumed need (even before the need crystallizes). A "research problem" offered by a scientist can, therefore, be a "solution" (in March's terms) to an organization problem.

Similarly, demand by the organization can be both March's "problems" *and* his "resources" (because I use the term demand to mean *real* demand in the economic sense). Consequently, though these two streams are tightly coupled in my modification of Zeldenrust's model, this only relaxes the requirement of "loose coupling" in one part of the model March proposed. In short, the March model is simply a more abstract version of Zeldenrust's and my model, albeit the terms (and one constraint) have different names.

Zeldenrust (1990) proposed his "garbage can" model because he noted that decisions about research problems to investigate were made in conditions of ambiguity about, for example, the problem's significance and even its technical feasibility. Zeldenrust also saw other similarities between the behavior of organizations choosing research problems and the decisions of the academic decision makers in universities and other educational organizations that March, *et al.*, refer to collectively as "organized anarchies." Like March's academic decision-makers, scientific problem choosers may offer solutions for problems that do not yet exist. However, in the case of scientists, the "solutions" are research problems and the "problems" are really external "demand." Just as solutions may seek problems without success, research problems may seek demand without finding it. And just as decisions in organized anarchies occur when the separate flows come close together and a linkage occurs between them, research problem choices occur when problems demand, resources, and (perhaps) constraints are linked together.

Because problems can seek demand (and vice versa, demand can seek problems) without success, problem choice cannot be "problem driven" if Zeldenrust is correct. This puts his perspective at odds with the rationalist perspective which presumes demand (at least for significant problems) and which, I also maintain, expects competition among scientists to solve these significant problems.

This study looked at five areas where the rationalist, social constructivist, and "garbage can" perspectives of problem choice in science differed in their predictions: (1) autonomy of scientists in problem choice; (2) local versus theoretical origination of problems (3) the relationship of competition/conflict to problem choice (4) organization culture as an influence on problem choice and (5) gender as an influence on problem choice. I contend that given the evidence adduced in this comparative study of the three perspectives, the "garbage can" model is the only satisfactory one as a basis for an empirical theory of problem choice. The rationalist model seems seriously wanting; its assumption of autonomy is open to question. Furthermore, its implication that there will be a considerable amount of interpersonal competition as scientists vie for the honor of solving a discipline's significant problems seems largely unsupported by data in my study or elsewhere (*e.g.*, Zeldenrust [1990] Mulkay [1991, originally 1974]). As shown in this study, the absence of competition may be because scientists are getting problems from many sources. They do not know the comparative theoretical significance of the problems, although they do know which ones interest their financial backers.

The social constructivist perspective likewise is wanting, albeit for different reasons. It cannot account for the considerable degree to which problems are not locally derived. It cannot explain such underlying regularities in problem choice as the influence of organization culture and gender. And the allegation (Latour and Woolgar) that scientists are driven by a "mania" to write appears farfetched; indeed, when one examines scientists at commercial research centers where few publications are done compared to academic research centers.

However, if the aim is to set aright the theoretical underpinnings of research into problem choice it is not sufficient to detail the deficiencies of the current perspectives on empirical grounds. If scientists are to reject these current models, it is necessary to offer some other more plausible model to take their place.

I maintain that the "garbage can" model, as modified here, can meet that requirement.

This is exemplified by its ability to throw light on gender based differences in problem choice. I think this phenomenon is more than a mere curiosity; indeed, I regard it as a strategic research site for evaluating models of problem choice.

To begin with, the social constructivist model cannot explain gender based differences in problem choice. This tradition of scholarship sees problem choice as "chaotic" and "local" and does not even imagine underlying regularities in the process. Thus, as a model of gender differences in problem choice it is entirely unsuitable.

The rationalist model also does not work well for the purpose of explaining gender based differences as shown in this study (see Table 5.1).

Gender based differences, according to rationalists, should not arise if scientists, as assumed by the model, have autonomy of problem choice. They should only occur when there is some interference from nonscientists, *i.e.,* a barrier(s) of some sort that differentially impacts scientists of one gender in trying to solve the significant problems of their discipline.

The rationalists are correct that women experience more interference, less autonomy, than men scientists do. However, there are limits on the autonomy of scientists regardless of gender. Rationalists seem intent on looking at such limits as inherently bad, as inherently having a negative impact on the quality of science. This is because the rationalists presume that scientists know what the important problems are. However, this presumption is not warranted. All we can be certain of is that scientists before they commence a study have a good idea of what the demand is likely to be for the results of their research. If scientists always knew more than this, it would be impossible to explain why scientists often *welcome, even seek*, direction from nonscientists (backed by resources, of course) regarding what problems to study. Far from limits on their autonomy being bad, some limits on scientist autonomy actually are beneficial for science.

An important consequence of the fact that scientists do not have perfect information about significant problems beforehand, that they need a clue to what will be desired by non-scientists who must furnish the resources for research, is that scientists can find problems not just in the paradigms of their disciplines but also inductively. That is, they can be guided by values and ideas differentially salient for each gender and for the embedding organizations in which they work. That gender (and also organization context) can play a positive role in problem finding is something that the rationalist model totally misses because it believes that demand is inherent in the problems and these important problems are known by disciplinarians beforehand based on theory commitments.

The "garbage can" model, however, would not blind the researcher to the role cultural variates play because it simply argues that problem choice is a linkage among four independent or loosely coupled streams: demands, problems, resources, and constraints. (In my adaptation it is a linkage between real demand, (which include resources) and problems, and sometimes constraints.) The "garbage can" model takes no position on where the problems are found. It also does not consider problems from one source, such as theory, as somehow better (or worse) than from another source. All it does is emphasize that problem flows and demand flows are independent (or loosely coupled). It is consistent, therefore, with a view that in addressing certain demands (*i.e.,* in wanting to study particular problems), scientists are expressing a *variety* of commitments -- commitments to values and ideas they hold dear as disciplinarians, perhaps, but also perhaps commitments to values and ideas they hold dear as members of particular organizations, or even different genders.

One implication of this point is obvious: after barriers in science differentially impacting women fall, gender based differences will still occur in problem choice, indeed may well be magnified if my data are correct. This result is quite difficult I think for the rationalists to explain. If for no other reason than that the "garbage can" model accounts more parsimoniously for gender based differences, it is preferable to any other currently available model of problem choice.

Now I will briefly recapitulate the evidence for these arguments.

Autonomy

It is difficult to overstate the importance of the assumption of professional autonomy to the rationalist model; without it, the model seems scarcely possible. Yet, prior research using a variety of study populations and research methods (Shenhav [1985], Nederhof and Rip [n.d.], Westfall [1977] and Zeldenrust [1990]) had strongly challenged this assumption's empirical basis. In my own research, based on data from a non-probability sample, (heavily biased in favor of senior scientists) I could muster only weak and inconsistent support at best for it. As indicated in Table 5.2, although most claimed to be free to study problems they wanted to, a large minority of researchers stated they had been told what to do by their research sponsor. And a majority of those answering the question stated that they would study a problem brought to them by superiors.

One possibly interesting finding on "autonomy" is the gender difference suggested in my data (See Table 5.2). However, the key finding of this section is that among scientists claiming relatively more freedom to choose gender based differences in problem choice were actually larger than among those claiming relatively less freedom to choose. This is the exact opposite of what the rationalists would expect.

My trumpeting of a finding that gender based differences are greater among relatively autonomous scientists is not intended to belittle the importance of constraints on women's research. On the contrary, in light of the findings of the "Study on the Status of Women Faculty in Science of MIT" (reported in "*The MIT Faculty Newsletter*" vol. XI, #4, March 1999), I think it is important to look at the gender based differences in scientists' autonomy[21] The possibility of gender patterned variation in autonomy raises basic questions about the organization of scientific research in North American research institutes (recall that my data include Canadian as well as United States based researchers).

At first glance, it seems odd that the rationalists, with their interest in structural issues, overlooked the possibility that teams headed by women researchers have less autonomy in problem choice than teams headed by male researchers (and that, therefore, external demand is more essential to understanding problem choices in scientific research than the rationalist model had supposed). However, Zuckerman suggests that scholarly attention may have been deflected away from this issue by the tainted image of externalist inquiry into problem choice. (As she says [1988: 547], the search for external influences on problem choice was a "not always respectable" line of inquiry.)

Regardless of whether my suggestive finding on gender patterned variation in autonomy is substantiated in future studies, the data I collected adds to the accumulating body of research showing that a theory of problem choice beginning with the assumption of researcher autonomy is of limited empirical value. The "garbage can" model, a perspective that does not assume researcher autonomy is certainly superior to the rationalist perspective in this regard. (Also superior from this standpoint is the resource dependence perspective of Shenhav (1985), an hypothesis entirely built around the idea that researcher autonomy is questionable. However, it does not address any other issues.)

Local Derivation of Problems versus Theory Derivation

If the findings on professional autonomy do not fundamentally shake confidence in the merit of any of the three models, the same cannot be said about the findings regarding whether problems chosen for research are derived from theory. If my understanding of the social constructivist perspective is correct, then my findings are a virtual refutation of one of the central ideas of the social constructivist tradition, namely, that problems are always locally originated. Outside of biology, where most studies in the social constructivist tradition have been done, there is hardly a field where local origination seems even particularly frequent, according to my data.

Yet my findings about derivation of problems can offer only scant comfort to committed rationalists. First of all, while local origination is not the main way problems are found it is not a rare method of identifying them either: in every field scientists obtain problems at least sometimes by local origination and in some fields it is quite common as suggested by Zeldenrust's (1990) archival research in Netherlands biotechnology research institutes.

Second, my data suggest that *absence* of theory, or a *gap* in theory, is motivational for scientists (to want to offer problems) to a far greater extent than suggested by Zuckerman's remarks that scientists choose the best problems (if free to do so) based on theoretical commitments or methodological commitments except in a few cases.

A second important finding in this section of the study was that whether a problem was theoretically derived or not will influence the probability of it being a "shared" problem (since problems chosen after comparison with other problems probably come from a stack of "shared" problems).

This finding has implications for the rationalist hypothesis. Rationalists assume that problems are "shared" among specialists. To the extent problems are not shared, however, they are less likely to lead to competition for priority in solving them.

The model most consistent with my results on where problems came from is the "garbage can" model. In neither the original March, *et al.*, version nor in Zeldenrust's does it make any predictions at all about where scientists will find the problems that ultimately become linked to resources, *etc.* when a problem choice is made. In this regard, it is superior to the social constructivist models which make the *wrong* prediction on this point. It is also superior to the rationalist model, I think, because the latter has a difficult task explaining why absence of theory is a motivator for many scientists.

Despite the fact that my data together with prior studies cast strong doubt on the value of both the social constructivist and rationalist models, I recognize that the evidence adduced thus far is not sufficient as a basis for abandoning either of them and embracing a different perspective. However, I believe that the necessary additional justification for embracing the "garbage can" model is found in an examination of findings from my study on interpersonal competition, organization culture, and gender as factors in problem choice.

Competition

In my opinion the rationalist model implies that interpersonal competition should be common and should play a role in problem choice. This is because the rationalists claim there are only a small number of highly significant problems that will attract talented researchers with an instinct for the truly important issues. These talented individuals are willing to endure (or perhaps have a thirst for) the keen competitive struggle for priority. People of lesser talent, or with no appetite for the rigors of such competitive struggle, will have to settle for working on the less important issues, according to Zuckerman (1988:83).

Zuckerman has stated the rationalist position on the importance of competition so persuasively that even some critics of the rationalist model endorse her position on that issue. Witness the comments of Mulkay (1991; originally 1974) and Zeldenrust (1990) cited earlier in this study about competition in science (see Chapter Two *supra*).

I believe that part of the reason for such wide agreement that competition is important in problem choice is that there are definitional problems with the concept of competition. At times, it seems that writers are using the term to refer only to interpersonal competition while at other times it seems that they may have in mind interpersonal competition arising from structural competition. No one can seriously question the influence of structural competition, which directly affects the ability to marshal the resources needed to address problems. But can the same be said about interpersonal competition? If prior research by Zeldenrust (1990) and Mulkay (1991; originally 1974) is any indication interpersonal competition may not be common.

I offer two kinds of data that seem consistent with these findings of Zeldenrust (1990): (1) data on whether there is reason to suppose that scientists are deriving their problems from a "stable" problem set and (2) data bearing more directly on competition as an influence on problem choice in science. I would like to comment on the data regarding a stable problem set first.

The results of my inquiry suggest that the assumption of a stable problem set of already known problems is not supportable. As evidence, I have answers respondents gave to an open-ended question on whether they knew and could describe briefly their next research problem.

Only a small fraction of the researchers in my study group could name their next problem or describe it. Despite some problems with the indicator, I believe that a much larger proportion of the study population should have been able to name their next study if problems were shared or known in advance by all members of the specialty. My assessment of this finding is that specialties with stable sets of known problems must be unusual, if not rare, in science.

Data on interpersonal competition I reported above demonstrate that scientists seldom get involved in such rivalries. As one scientist put it, "It's a waste of time." Still, more than half of my respondents (54%) from the exact sciences admitted that they were motivated to beat out a rival at some point in their careers. This suggests that the sense of interpersonal competition was most likely engendered by structural competition. There was demand for the results of a particular research project and other researchers were aware of this demand. It was important to be first in this race. But it was not the interpersonal rivalry that was primary. It may even have been epiphenomenal since it neither preceded the research not continued beyond the project finish.

When the embedding organizations are engaged in a struggle of oligopolistic competition in an industry where success means huge profits, it is possible that the rivalries engendered by such competition breed more durable interpersonal competition among team leaders of rival company research teams. Thus, interpersonal competition may be more common in the pharmaceutical field and a few other specific areas than it is in general. But as I have insisted, this is because there are enormous stakes, both financial and otherwise, in making significant scientific breakthroughs in the pharmaceutical industry. (For example, discovering an effective chemotherapy for common killer cancers such as cancer of the lungs could generate billions of dollars in sales and huge profits for the company owning the patent.) Thus, there is a structural basis to the rivalry. Still, the top scientists working on the major problems in pharmaceutical research undoubtedly know of each other, and may even be acquainted personally, thus giving rise to the conditions needed to foster interpersonal competition and conflict between them.

I have now addressed much evidence for believing that the rationalist hypothesis is erroneous. First, I have shown that scientists are not likely to be autonomous in problem choice (See Table 5.1). I have also shown that problems can come from anywhere (See Table 5.3). Furthermore, where the problems come from influences the degree to which they are likely to be "shared" (See Table 5.5). Finally, consistent with these findings, I have shown there to be only limited interpersonal competition and limited evidence of the "stable" problem set I would expect if scientists were competing to solve the most important questions.

My conclusion is that the rationalist hypothesis is correct for a limited set of circumstances. *When* there are "shared" problems and scientists are relatively free to choose, they will compete to solve the most important of the problems. This might be an accurate characterization of a few fields such as high energy physics or astrophysics. It probably does *not* characterize many other disciplines or even some other specialties in physics itself.

I regard the social constructivist perspective as flawed. Its main contention that problems are locally originated and the process of problem choice is chaotic is simply erroneous. Most problems are *not* locally originated in many fields and while problem choices cannot be predicted, *characteristics* of the problem choice are predictable.

Before leaving the topic of competition, I want to add that some interesting findings demonstrated gender differences in desire for priority in competitive situations (Table 5.9). These findings suggest that women may be more likely to shy away from interpersonal competition than men. Since Zuckerman (1988) suggests that competing to solve the most important problems of the discipline involves head-to-head competition (in certain hot fields, for example) and Cole and Singer (1991) suggest that there may be limited differences between men and women that account for the gender gap in scientific productivity--an area where there has been considerable research in the past two decades—these findings about gender differences in competition may clarify puzzling findings of other scholars. I shall have more to say about this in Chapter Six below.

However, returning to the main point of this summary discussion of competition, it should be clear from the evidence of this and other studies that the two main models are wrong about competition albeit for different reasons. On the other hand, the "garbage can" model makes no assertions about competition and does not imply that competition should be common. Indeed, given its intellectual debt to March's views about decision making in conditions of ambiguity, it is difficult to believe that the "garbage can" model would argue that typically scientists will identify similar problems and marshal the resources needed to address them, especially in disciplines, such as macro-biology, that lack the powerful paradigms common in physics. Rather, Zeldenrust's model would lead its adherents such as myself to argue that (a) scientists are interested generally in working on problems that they can find some demand for solving as evidenced by the willingness of funding sources to pay for the research and (b) scientists may have no really clear idea of how important their ideas regarding particular research projects are until they are quite deeply involved in performing the research -- and perhaps not even then. And quite possibly, they do not have a full understanding of, nor do they care greatly about, the implications of their research until perhaps long after it has been completed and further research has been done based on their findings.

Just as interpersonal competition seems to be unusual (outside of a few specified areas listed above) and hardly important as a factor in problem choice, conflict at the personal level is likewise not common and not a significant motivator of problem choice. It is probable that structural competition, *e.g.,* imperfect competition between organizations, however, does have a direct impact on problem choice. There also is probably an indirect influence because inter-organizational competition, a form of structural competition, may influence the scientific group's work culture. That influence is reflected when the scientist demonstrates commitment to the institutional culture through studies she does supportive of the institutional culture's chosen paradigms, or favorite methodologies (*e.g.,* computer modeling) or both.

Organization Culture

The most distinctive contribution of this study, in my opinion, is its demonstrating that problem choice results from a cultural process that is nonrandom and non-rational. The process does not guarantee problems to be “significant” or “not.” Sometimes the process has positive effects as when it encourages scientists to seek out intellectually challenging problems rather than problems that can lead to easy publications. Alternatively, the process can lead scientists to reject worthwhile research in favor of technically demanding work on minor issues.

It is difficult to exaggerate the importance I place on seeing problem choice as a result of a nonrandom, non-rational cultural process at the organization level. In making my case for this point of view, I proceeded along two complementary paths. First, my findings on autonomy and theoretical versus local derivation of problems cast doubt on the adequacy of the rationalist model and the social constructivist perspective. Second, once it was clear that those two models were deficient in certain respects, my findings on the role that organization variates play in problem choice (see Tables 4.9 through 4.12) clearly showed that the March and Zeldenrust formulations are better than any currently available rival models. This is because the “garbage can” model offers a satisfying explanation of (1) why the organization wants to socialize scientists to offer relevant problems and (2) why organization, not the scientists themselves, select the problems studied although the scientists may have a large role in the process in many cases.

Zeldenrust never considered a role for organization culture in his model. Instead, Zeldenrust had tried to use the concept of "driving force" which would be different from one work group to another or over time within the same work group. Thus, in one work group the "driving force" would be demand, while in another it might be problems. Zeldenrust also shows that the driving force could change as key members of the research team departed and were replaced by others with different research orientations and professional role expectations.

It seems to me for a number of reasons, however, that organization culture is a more satisfactory concept than "driving force." For example, the concept of organization culture helps in understanding why (a) sometimes "demand" would be a primary concern and "problems" would be sought to address the "demand" and (b) sometimes "problems" would be the primary concern and "demand" would be sought to support work on those problems.

The most important reason I emphasize the role of organization culture in problem choice, however is that it underscores my thesis that problem selection in science is not "problem driven." I can think of no better evidence of that than the comment of one of my respondents (SO55) who remarked that the most profitable problem *technically* is not the one he would necessarily pick given the culture of the organization in which he works.[22] He would choose a problem that was technically demanding to solve though of marginal intellectual significance over a more intellectually important issue that would not be technically demanding. This is because he must take into account his research organization's strong need to differentiate itself from other research groups elsewhere in the academic or non-academic world – *i.e.*, his organization of orientation's need for boundary maintenance in his problem selection. And he is not likely to be unique in this regard.

In developing my thesis that it is the organization culture of the research team that is central to the problem choice, I made a number of specific points based on findings of my study.

The first of these theoretically significant findings about the importance of organization culture is that (a) the culture of the work group would reflect some of the embedding organization's values, and (b) this would in turn be reflected in the problem choice of the team. Especially apposite in this regard are the results of Table 5.14 (p.198 *supra)* regarding a relationship between acquiring special equipment (e.g. magnetic imaging machines) and "the organization's desire that a novel analytical approach be used" in the research project.

Though it seems almost obvious, it is necessary to point out here also that my research shows that scientists in bureaucracies are mindful of the political climate affecting their employer and the essential constraints it imposes on their problem choice. My data showed that biological researchers, for example, avoid experiments with animals of interest to powerful (and often militant) lobbying groups even if the research might be highly beneficial. And news accounts have indicated that because of religious opposition and legal restrictions, scientists are avoiding doing fetal tissue research in the United States.

The integrating of some of the values of the embedding organization into the scientist's values and its reflection in problem choice would come as no surprise to a Parsonsian (structural-functionalist). It is necessary for the scientist to incorporate these values to facilitate organizational integration. By making this point, however, I further undermined the rationalist model, while enhancing rival models that allow for a role for external demand, *e.g.,* the resource dependence perspective.[23]

One caveat is in order here. While the cross tabulations on organization trait effects on problem choice criteria illustrated how organizational imperatives are taken into account in researcher decisions on which problems to investigate, I do not argue that the organizational imperatives necessarily lead the scientists to choose problem choices most agreeable to the organization. I merely claim that the researcher *assesses the implications* of the organizational imperative before plunging ahead on a problem. For example, the researcher may do a study that utilizes a fancy mathematical model because she sees this as necessary to meet organizational expectations. Or she may forego using ideal animal subjects. What must be remembered here is the element of calculation, of planning in the problem choice process, that Knorr-Cetina seems to strenuously deny in her claims that research decisions are merely opportunistic, not rational, and not predictable (see Zuckerman [1988:555]).

However, I did not find the degree of influence that Shenhav (1985:48) expected to result from resource dependence upon others when he said, "external control can force researchers to work on certain research areas rather than others, can bring them into formal groups or projects, can constrain their available time for 'free' research, and can disturb the process of information flow."

It is important to remember, of course, that I did not try to replicate Shenhav's study by utilizing any of the instruments he relied upon for his analysis. Therefore, I simply do not know to what extent problem choice is constrained by external influence generally.

It should be clear now that scientists do take account of their organizational settings in their problem selection; that problem choice is not simply problem driven or demand driven as the extreme rationalist and resource dependency models would assert.

Gender

My claims for the importance of the cultural dimensions in a theory of problem choice in science are most strikingly demonstrated in the case of gender's role in problem choice. (As stated earlier, gender is a cultural variable, not a biological one.)

Women scientists weighed certain criteria differently than men. For example, women were more likely to value the importance of legitimization by senior scientists before starting a project (See Table 5.19). This is clear from the fact that women reported seeking such legitimization before starting more than men scientists did. Perhaps this finding reflects the lesser status of these women in academia, *etc.,* and their perceived need to get prior approval. Whatever the reason(s), the differences I found between men and women were substantial.

The most dramatic data is support of my thesis that the "garbage can" model is superior to any other current model of problem choice was the evidence adduced in Table 5.2A and 5.2B showing that when professional autonomy is high the relationship between gender and problem choices is strongest. Rationalists would predict that when autonomy is high, gender based differences in problem choices would disappear.

The last point I want to emphasize here is that gender differences are not merely interesting in their own right; these differences may bear on the puzzling finding in this study and elsewhere of a gender gap in productivity. I pointed out, for example, the greater attention that women scientists devoted to buttressing the legitimacy of their work by first seeking the approval of respected scientists for a new project.

This kind of care women scientists gave to marshalling symbolic as well as real resources may be a factor in their somewhat lower productivity scores compared to men scientists. But other factors may also be at work; for example, I showed that, compared to men, women reported more interference by institutional controls, forcing them to alter or drop planned research.

These data taken together show the limitations of models of problem choice that emphasize the rationality of choices and also of models that seem to regard the process as chaotic. In short, the data underscore the superiority of a model that forcefully makes clear that scientists are making decisions in ambiguous circumstances, bringing to bear knowledge and values to make a "best" choice in the circumstances. This view will probably not be easy for some scientists to accept. It is comforting to see problem choice in science as a "rational" activity; a mental labor that is more exalted than ordinary thinking. It will be difficult to shake people from this view of the special nature of arriving at a problem choice.

Yet, I think there is a repellent quality to the rationalist model's image of scientists as passive, cool appraisers of problems associated with the received theory. In fact, some sociologists seem to have found this image so unattractive that they were ready to embrace the social constructivist perspective. If nothing else, Knorr-Cetina, B. Latour and S. Woolgar, H. Collins, and others associated with this view have put the human animal back in charge of problem selection. However, the animal appears to be motivated by a "mania" to write (see Latour and Woolgar [1979]). Research problems in the social constructivist perspective seem to have no relation to core ideas of the discipline, perhaps because for the social constructivists such core ideas do not exist. (see S. Cole 1992). As depicted in the research and theorizing of the social constructivists scientists who identify problems, are opportunists who originate their problems locally, with the help of machines and other unique environmental features of the laboratory where the scientific work is done; their decisions in this regard are (in the words of Zuckerman [1988:555]) "not rational, well planned or rule governed," In its more extreme versions (for example, H. Collins) the social constructivist approach denies any reality other than a subjective one and cannot avoid the philosophical cul-de-sac that debates about what is real and how can it be known usually degenerate into. In other versions, such as recent statements of Knorr-Cetina, it seems hardly different from the rationalist perspective from which it is at pains to differentiate itself (see Zuckerman [1988:556]). From my standpoint, it fails to account for the high degree to which problems are theory derived -- an important failing in a perspective that insists on predicting where scientists find their problems. And, like the rationalist model, which is problem driven, the social constructivist perspective does not allow for such underlying regularities as the influence of gender.

That sums up my quarrel with the rationalist and social constructivist views. Their images of scientists are too extreme: either too passive and analytical (rationalist) or too manic and too opportunistic (social constructivist) in finding problems. A model that recognizes that scientists are active players, but that also recognizes that they *usually* get their ideas from the core ideas of their discipline seems more plausible than either of the other two views. The "garbage can" model is that model. Alone among the models reviewed it finds the middle ground.

Technical Appendix -Chapter Five

The Index allows various combinations of events to be counted as "rational" problem choice:

1. Problem chosen after Literature Review
2. Problem chosen after Renowned Researcher Urges Work on it.
3. Problem chosen after *Comparing* Two or More Problems Based on Technical Criteria.
4. Problem chosen after a Sponsor Demanded Work on it, *if* either #1, #2, or #3 are *also* present.

Choices made that are "non-rational" would not meet *any* of the criteria 1-4 given above.

Based on these criteria, I found 80 "rational" problem choosers I in the 104 cases for which I had usable data (see below).

Goodness Of Fit Calculation - 95% criterion

	Inexact Sciences	Exact Sciences	Total
Observed (N of Cases)	43	37	80
Expected (N of Cases)	55	44	99
Total Cases	58	46	104

$X^2 = \sum \frac{(fe-fo)^2}{fe} = \sum \frac{(55-43)^2}{55} + \frac{(44-37)^2}{44} = 144/55 + 49/44 = 2.62+1.11=3.72, p<.05$

The rational choice (null) hypothesis is rejected at the 0.05 level.

Goodness Of Fit Calculation - 90% criterion

	Biological & Social (Inexact)	Other Sciences/ Engineering (Exact)	Total
Observed	43	37	80
Expected byRational Model	52	42	94
Total Cases	58	46	104

$X2 = \sum \frac{(fe-fo)2}{fe} = \sum \frac{(52-43)2}{52} + \frac{(42-37)2}{42} = 81/52 + 25/42 = 1.55 + .6 = 2.16, p<.05$ n.s.

The rational choice (null) hypothesis is sustained at the .10 level.

ENDNOTES TO CHAPTER FIVE

[1]Details of the Index construction and the rationale are contained in the technical Appendix to this Chapter.

[2]*See* Appendix.

[3]Generally only "shared" problems can be compared. An exception would be a stock of prior findings at a laboratory that have not been shared with other laboratories. I regard this as an unlikely situation in academic research (in classified research, possibly comparison among a stock of prior findings from the same laboratory might occur with some frequency. However, I did not investigate this).

Zeldenrust identifies T. Gieryn as a proponent, along with Zuckerman, of the rationalist model. I share Zeldenrust's opinion. Given that Gieryn studied astronomers' choice of problems it is hardly surprising that he is convinced that the rationalist model is adequate since in astronomy there are powerful models from which to derive problems. Gieryn, however, offered no evidence on the issue, since his research dealt with *specialty* choice rather than problem choice, as he himself pointed out (1978).

[4]This question was inadvertently omitted from some of the questionnaires.

[5]*See* also Zuckerman (1977;1988) regarding competition in science.

[6]This analysis is not shown in the study.

[7]Kuhn (1969: originally 1962) in various places makes statements suggesting that problem derivation is straightforward. Thus, (1969:166) he says, "In its normal state, then, a scientific community is an immensely efficient instrument for solving the problems or puzzles that its paradigms define." Again, he says (1969:107-108),

Initially, Maxwell's theory was widely rejected.

> ... But like Newton's theory, Maxwell's proved difficult to dispense with, and ... it achieved the status of a paradigm .... The result ... was a new set of problems and standards, one which ... had much to do with the emergence of relativity theory.

[8]It was surprising to me to see evidence of desire to beat out a colleague more frequently in the exact sciences than in the inexact sciences. I surmise that this is because given the existence of powerful paradigms in the former group of disciplines, proportionately more "well-formed problems in the exact sciences are

known to all disciplinarians than in the inexact sciences." Consequently, in the exact sciences there is keener competition to (a) obtain resources and (b) once resources are found, to solve problems before another team does. In the inexact sciences, finding a well formed problem is more difficult; but once such a problem is found by a scholar there are not too many others likely to be working on it.

[9]Merton further hypothesized that kinds and degrees of competition also differ among prestige strata, a point I did not pursue. However, I think it more probable that there is an *interaction* effect between prestige level and specialty so that even among prestigious researchers in certain fields, I would not expect to find much head-to-head competition while in others competition would be quite intense.

[10]Hagstrom looked at competition in the animal world and he saw (as Darwin put it in *The Origin of Species*) that:

> ". . . the struggle will invariably be most severe between the individuals of the same species, for they frequent the same districts, require the same food, and are exposed to the same dangers." In the case of varieties of the same species, the struggle will generally be almost equally severe, and we sometimes see the contest soon decided.

How can Hagstrom and Zuckerman reach such differing conclusions about competition in science? I believe that they are really looking at differing aspects of science. Hagstrom, looking at scientists generally saw that head-to-head competition on the interpersonal level could not be sustained for more than a brief period. This is why, for example, I believe that Mulkay's radio astronomers were quite correct in their assertions that they largely avoided head-to-head competition with other radio astronomers in the United Kingdom.

At the same time, of course, these same radio astronomers could be intensely interested in winning awards and receiving grant funds that are *competitively* distributed. But one scientist would be entering the prize competition with a study on quasars, for example, and another with a study on evidence for planets circling "nearby" stars in the Milky Way galaxy. To my mind, Mulkay did not give due emphasis to this division of labor.

In speaking of the intense competition in science, Zuckerman was certainly alluding to this institutionalized competition for various honors. Zuckerman's point is that Nobel Prizes, other prestigious honors such as the Lasker Prize, or the Max Planck Prize in Germany, opportunities to join the faculty of prestigious science departments, *etc.,* are all *competitively* distributed.

It is possible though less clear if Zuckerman also had in mind the head-to-head competition of people trying to do the same things (an example of the latter being Watson and Crick's race to beat Linus Pauling in discovering DNA's structure).

Regardless of what Zuckerman was referring to in arguing that there is "intense competition" in science, I do not completely share her views. Head-to-head competition is the exception *within a country* occurring usually only in the most financially lucrative areas such as new chemotherapies for treatment of cancer and heart disease. Therefore, this kind of competition would not be a source of the "intense competition" Zuckerman believed characterized the world of science. Structural competition for major prizes and honors might be a more likely cause of intense competition. Bear in mind, however, that many quite competent scientists, who publish voluminously, will never win a Nobel Prize, etc., for the simple reason that none is awarded for outstanding work in their particular field. For these scientists there is no "intense competition" for prizes. Indeed, I doubt that the vast majority of scientists would claim that there is intense competition. If they think about competition, they are concerned about broader forms of competition--between science *as a whole* and other claimants for federal dollars, for example. For academic scientists who are searching for grants, however, Zuckerman may well be correct. They are keenly aware that research funds on which they depend are increasingly hard to get.

[11]Not only may there be a "pressure to publish quickly" as Merton alleges but there may be a pressure to seek various other edges over the unknown competition. For example, I found that there was a strong positive relationship *among men only* between the "researchers wanting to choose problems allowing a novel analytical approach to be employed (V21R) and a perception that "other researchers are migrating into [the researcher's] specialty." Eighty-one percent (N=17) of those men respondents claiming that other researchers were migrating into the specialty were interested in selecting problems permitting a novel analytical approach to their data versus 46% (N=15) of men not claiming that other researchers were migrating into their specialty".

My data suggest that this may not be a gender difference. It may, in fact, be a disciplinary difference (or an interaction effect of gender and discipline). For example, among those respondents in "exact" disciplines 81% (N=13) of those who perceived that "other researchers are migrating into [the respondent's] specialty versus 39% (N=9) of those not claiming such in-migration indicated an interest in choosing problems allowing a novel analytical approach to the data.

Not only was the fact that it was in exact disciplines that the relationship occurred a relevant issue. Whether the discipline was in a "turbulent" state intellectually was also relevant. Thus, for example, among those who thought their specialty was in a turbulent state, 74% (N=14) perceiving an "in-migration" versus 35% of those not claiming an in-migration of other scholars stated that their problem selection would be influenced by the possibility of applying a novel analytical technique to the data.

[12]The respondent (S063), a renowned mathematician, was embroiled in a controversy over his work. He told me in an interview:

> S063: People resist my ideas
>
> Int: Why do you believe that to be?
>
> S063: My critics don't want to hear about my work. They have an ideological fear . . . They [my critics] never invited me to debate with them [because] they know I'm smarter than they are.

[13]Seventeen percent of the 87 respondents, answering this question thought this occurred at least sometimes. The vast majority disclaimed any knowledge of such activity or were certain it was rare or never occurred.

One of the few scientists who thought that some scientists "choose problems for research primarily to embarrass their rivals or personal enemies" - she wrote that "yes [it] occurs very often in my area" - is S022, a pharmacologist working in the commercial pharmaceuticals area. Her divergent opinion can perhaps be explained by the fact that it is well known that successful drugs can result in billions of dollars of sales revenues for pharmaceutical companies. Attacks from scientists working for rival manufacturers on researchers who help to establish the documentation of the efficacy and safety of such potentially lucrative substances seem to have a logical basis in inter-organizational conflict arising between oligopolistic competitors in the pharmaceutical industry. Given competition for this extraordinarily lucrative business, the pecuniary motive seems perhaps to overshadow any norms of not being too negative in one's comments about a fellow scientists' work (if that fellow scientist works for a competitor or has contributed work enhancing the competitive position of the other firm).

[14]It is ironic that the unit established to do research leading to an understanding of some phenomenon instead would focus on research studies whose primary purpose is to differentiate the unit from other research units elsewhere. However, this displacement of goals is a common phenomenon in organizations. Perhaps it is inherent in organizations to move in this direction. White (1992) seems to suggest as much. In his discussion of "blocking action," White (1992: 147-50) points out that at the organizational level it occurs concomitantly with the formation of organized units able to carry out the needed work -- *i.e.,* needed from the standpoint of unit productivity (or rational problem choice), *etc.*

White (1992:149) gives a telling example from the economic history of banking in England based on Fine and Harris (1983).

> The banking system that was constructed (by 1914) put the banks in a position where they acted as a block against the external forces that were necessary for industry's growth, and it has been the structure of the system rather than the attitudes and

> choices of bankers that has been at the root of the problem . . . The banks have met industry's demand for credit all too comfortably, and, as a result, have developed a special relationship with industry which has given them a blocking role.

Just as in the example White gives regarding the role of banks in exerting a blocking action, there is a blocking action exerted by the organization presumably dedicated to first rate scientific work. The mechanism is the peer review system of science. The scientist needs her colleagues to pass judgment, to accord the desired *recognition*. In turn, to get professional recognition - "our applause" as economist Paul Samuelson phrases it - she passes up the technically most significant problems in order to do work that ensures the distinctiveness of the scientific organization, *e.g.*, the university chemistry department or the government laboratory, *etc.* This displacement from the intellectually most significant problems is a most profound example of the way in which the influence of organization culture is (negatively) manifested in problem choice.

[15]This point can be argued more formally from a network theory point of view. For instance, Granovetter (1977; originally 1973) might notice that principal investigators have ties to the organization superiors who themselves are not researchers. Granovetter (1977:349; originally 1973), posits that "the stronger the tie between [the principal investigator] and [the organization superior who is not himself a researcher], the larger the proportion of individuals in the [organization] to whom they will both be tied."

The point is that, given all these ties, organizational influence on researcher values, perceptions *etc.* will flow through the ties and thus network theory would suggest why Caplow's point is, indeed, correct.

[16]Unfortunately, I did not explore organizational set in my questionnaire although its importance now appears manifest. There seems to be no comparable concept in network theory to "organizational set" as used by Caplow (1964:) and others working in organizational studies. White (1992) for example, was fully aware that people and organizations engage in comparative assessment but otherwise did not address the distinctive functions of the set at the organization level. He says, (1992:38) a set of actors can become comparable, become peers, through jostling to join in a production on comparable terms. They commit by joining together to pump downstream versions of a common product, which are subjected by them and downstream to invidious comparison . . . . Manufacturers of recreational aircraft for the U.S. market . . . can be examples [of this phenomenon].

[17]The Fisher's Exact Test result of .1 for a two-tail test is indicative of borderline statistical significance.

[18]Cole (1979:68) considered various possible explanations of the gender

gap in scientific productivity. He was able to show that several popular ideas about the cause of the difference were not supported by the evidence. Clearly uncomfortable about the gender gap he suggested

> . . . in order to account for the correlation between sex status and the rate of scientific output it may be necessary to look outside the institutional structure of science, to examine carefully the prior experience and socialization processes affecting women in the larger society which may dampen motivation to succeed and influence their publication performance after they enter science.

Although, I accept the possibility that a search for an explanation of the gender gap in events of many years earlier may ultimately prove necessary, I do not believe that the time to pursue this is at hand. We should first seek simpler explanations based on events more proximate to the productivity gap, e.g., in the process of problem choice and of research itself as performed by men and women scientists. Surprisingly, however, while sociologists (e.g., Cole and Singer [1991]; Fox [1995]) have offered numerous possible explanations of the gender gap, no one has put forth an explanation based on careful analysis of the research choice process itself. Apparently no one has even suggested that part of the answer to the gender gap may lie in the problem choice process.

The present study based on small numbers (especially of women) can not yield the definitive answer, of course. A further hindrance to comparing my results and Cole's 1979 study is that my research population is not exclusively a group of academicians as is Cole's. (However, to the extent more men than women work in industrial research settings, this would *reduc*e the gender gap in productivity based on *publications*.) This is because, as Shenhav (1985:137) remarks, . . . the higher the amount of external [to science as an institution] funds, the lower the number of publications. He adds that "Higher amounts of external funds reflect higher external control and require accountability to the funders who may not have an interest in the production of publications."

[19]These latter results may be somewhat affected by age differences (years at risk) between men and women (men are older on average in my study population). I did not examine this issue but the *possible* influence of differential risk exposure is acknowledged. Age difference should not affect results in the first table but *generational* differences as opposed to gender differences may be at work there.

[20]The analysis reported here is a replication done with Dr. Lawrence Lessner of an earlier analysis by the author. The main difference between the two analyses is that the present analysis contains a consideration of the best model of each of the various dependent variables. Fisher's earlier analysis only attempted to identify the best predictors and did not consider the best model. While in the

present study the best predictors were found without needing to determine the best model this is often not the case. Usually, depending on the model chosen, the predictor set will differ; therefore, determining the best model is usually necessary to determine the best predictor set.

[21]This is because the MIT study showed that

> ... in some departments (in the School of Science, one of the divisions of the MIT) men and women faculty appeared to share equally in material resources and rewards, in others they did not. Inequitable distributions were found involving space, amount of 9 month salary paid from individual research grants, teaching assignments, awards and distinctions, inclusion on important committees and assignments within the departments. While primary salary data are confidential and were not provided to the committee, serious underpayment of senior women faculty in one department had already been discovered and corrected two years before the committee formed.

CHAPTER SIX: THE RELATIONSHIP OF DISCIPLINE TYPE, ORGANIZATION CULTURE, AND PERSONAL PROBLEM CHOICE PREFERENCES TO PRODUCTIVITY

Prefatory Remarks

In the Introduction to this study, I asked, "Are there gender differences in the problem choice process?" I contended that if research problem choice and scientific productivity are related issues, I would find gender differences in problem choices since numerous studies show gender differences in scientific productivity and the quality of research publications.

The present study has been largely concerned with marshalling evidence, both from the literature and from my own survey data, that there are indeed gender differences in the problem choice criteria of scientists. One task remains: although I have asserted that scientific productivity and research problem choice, each a mature specialty within the sociology of science, are connected, I have not developed this point. In this chapter, I will take up that connection.

My aim in this chapter is broader than simply showing that the research problem choice process and productivity are related. For example, I will show how differences in the research problem choice process among disciplines influence relative productivity. Discipline, in a sense, is an organizational variable because membership in or affiliation with a discipline implies having an association with others of that discipline who are influential in your views about what constitutes good science and so on. I will then turn to other organizational influences on productivity and then to personal research styles and problem choice preferences that may impact on productivity.

I will begin by proposing a new hypothesis of disciplinary differences in productivity based on whether the discipline has strong or weak paradigms. This hypothesis—which I refer to as the "rationalist hypotheis of productivity" is distinguished from the rationalist perspective of problem choice identified with Zuckerman, Gieryn, J. and S. Cole.

After stating the hypothesis, I will provide a theoretical justification for each component of it. This will then be followed by evidence from the present study bearing on the hypothesis. After that, I will present conclusions and suggest further directions for research.

Theoretical Orientation

I contend that the type of discipline and other organizational variables are important in understanding scientist productivity. While I will reserve a formal statement of the hypothesis until later in the chapter I will say here that my position is that in disciplines with strong paradigms, specifically physics and chemistry, there is greater researcher autonomy; as a result there will be more competition for priority among scientists; and therefore scientists will have higher productivity. In contrast, in disciplines with weak paradigms-- for instance, several social sciences and some branches of medicine and biology—there will be lower researcher autonomy, less competition for priority, and lower productivity.

Research Productivity: Its Causes and Consequences as Identified in the Literature

The research productivity of scientists has been a central focus of social science research for the past forty years with much of the attention directed at gender differences in publication rates. Considerable evidence amassed over several decades points to a gender gap in scientific productivity. Valian (1998), summarizing the literature, reported that women published about half as many papers as men (p. 262)." The gap remained through the 1990s, although it appeared to be narrowing. (See Zuckerman [1987]; also Valian [1998]. Research carried out in the mid 1990s on economists and sociologists show no gender differences in productivity within those two disciplines (see Singell [1996]; also Stack [1994]). However, Valian points out that "taken together, studies measuring productivity in terms of quantity find that women typically publish 50 to 80 percent as much." (Valian [1998, p. 262]).

Two general perspectives dominate in this area: a differential resources perspective and a "limited differences" perspective. The differential resources perspective, championed by such scholars as Mary Frank Fox (1992) and, more recently, exemplified in the work of Barry Bozeman (forthcoming) points to the more limited resources women work with when they do research.

Bozeman and Lee (2004) reporting the results of a large survey they carried out, stated that "the number of collaborators is the strongest predictor of a scientist's productivity—as measured by books and scholarly papers published." However, he found that women have fewer collaborators (Bozeman [forthcoming]). Other scholars have noted that women receive less funding for research than male colleagues (Feldt[1986]); or are given less advantageous and/or more committee assignments (M.I.T. [1999]. The MIT study (1999) also found that in some departments at that renowned center of scientific research women faculty had more difficulty in obtaining student research assistants than men faculty did. Other scholars in the "differential opportunity" camp emphasize that fewer women are in senior faculty positions than men while proportionally more women are in junior and non-faculty positions (see Xie and Akin [1994]; also Xie and Schauman [1998]). Valian (herself emphasizes that women are adversely impacted by their lesser representation in the elite research universities. She remarks (1998:263) that "the culture at prestigious universities does not just help scholars be more productive than they would be at less elite institutions, it forces them to be."

The second main group of scholars argue that it is not a lack of resources that causes the differences in productivity but rather a difference in productivity that leads to differences in resources for research. For example, in a mathematical elegant statement of the argument Cole and Singer (1991) assert that it is the lower productivity of women that results in their disproportionate representation in non-elite universities and in lower faculty ranks of institutions of higher learning.

Cole and Singer (1991) attribute the lower productivity of women to "limited differences" that arise in the early education of future

The argument can be illustrated in the following example. Women do not enter the higher professional ranks proportionally as often as men do because they do not get into the junior positions as often as men. And the reason for that is "limited differences", *e.g.,* that proportionally fewer women than men pursuing graduate studies are supervised by senior professors and proportionally fewer of those women who are supervised by senior faculty land appointments as junior faculty at prestigious universities. (See Cole and Singer [1992], esp. p.288-297).

Proponents of the Cole and Singer position would also argue that fewer women graduate students are supervised by prominent men faculty because fewer talented women undergraduates are encouraged by their college professors to seek graduate training at the prestigious graduate schools from which junior faculty at elite research institutions are recruited.

Valian believes that the productivity differences between men and women reflect both differential resources and "limited differences" in socialization—a view that I generally share.

Cole and Singer's hypothesis (1991), though seductive, is untested more than a decade after it was advanced. .This is not surprising. Despite its intuitive appeal, the Cole and Singer hypothesis is fiercely difficult to test as a practical matter since only an enormously expensive longitudinal study could assess it adequately. Valian's own hypothesis emphasizing organizational effects as well as socialization (a nod to Cole and Singer) is likewise unproven for the same reasons..(Since it argues that "organization culture" and socialization differences are behind the gender gap in productivity, to the fierce problems of detecting socialization differences must be added the difficulty of operationalizing the concept of "organization culture."

Xie and Shauman's (1998) conclusion that an adequate explanation of gender productivity differences remains elusive is still correct.. Indeed, fierce debates about the causes and effects of these gender differences continue to rage.

While gender differences in productivity have received the lion's share of scholarly attention, some work has been done on other issues in scientific productivity. Recently, Dietz (2004:49) found " center affiliation differences (where the scientists and engineers tended to be more productive than in other centers" in an analysis of a survey of 450 scientists and engineers. Dietz's finding is the first convincing data suggesting the possible importance of organization traits such as organization culture in explaining productivity (for a discussion of "organization culture" see Sackmann, [1998]; also Chapter Five of this study *supra*).

Disciplinary characteristics may be a much more significant contributor to research productivity than organizational traits. Nearly thirty years ago Hargens (1975:61) noticed that chemists had a high productivity rate. However, this finding did not seem to generate much interest in identifying disciplinary characteristics that might explain productivity rates. An important exception is Dietz (2004:49fn22) who, in addition to organization differences, also explored disciplinary differences in productivity in his large sample survey. He reported that "… chemists, physicists, electrical engineers, and chemical engineers had among the highest publication rates." Furthermore, civil and mechanical engineers had lower publication rates than other kinds of engineers while general biologists had lower publication rates than other scientists.

Perspective of the Author

A new perspective is needed in the study of scientific productivity since none of the current perspectives is satisfactory. In this section I first state the proposed general hypothesis; then explain its advantages over current models and finally describe in some detail its rationale.

Hyp. 1 Scientific productivity is a function of

Gender;

Type of Discipline;

Professional Autonomy;

Willingness to Engage in Competition for Priority;

Organization Factors and;

Personal Problem Choice Preferences and Research Styles of Scientists

There is no dearth of hypotheses about the causes of the gender gap in scientific productivity but I maintain that there are two good reasons for offering a new hypothesis: (1) it attempts to explain scientific productivity rather than the more limited issue of gender differences in productivity and (2) it is testable with survey data.

It is important to focus on explaining the causes of productivity of scientists rather than on the "gender gap" in productivity. While there may be disagreements among them about the causes of the gender gap in productivity many social scientists concur that the gap may well disappear in the coming decades as it apparently already has in some disciplines (see Xie and Shauman [1998] and Valian [1998]; also see Singell [19996] and Stack [1994] quoted in Valian [1998]). For example, rationalists who attribute the gap to interference with the work of women scholars foresee a time when there will be no differential interference in scientific research. Other writers, such as Cole and Singer (1991) who see the problem as stemming from socialization differences, expect that in the future women will receive as much encouragement to pursue scientific and engineering careers as men. However, the larger issue of scientific productivity and its causes, especially to the extent they are policy variables that can be manipulated, will continue to be a focus of interest well into the future. Hence, there is a need for theorizing on the causes of scientific productivity regardless of whether there is a gender gap.

The testability of the new hypothesis offered here is equally compelling as an argument for giving it serious consideration. Despite its intuitive appeal, Cole and Singer (1991) is fiercely difficult to test as a practical matter. Only an enormously expensive longitudinal study could assess it adequately which no doubt helps account for the fact that it is untested.

Though I have outlined my general thesis in the prefatory section of this chapter, I have not stated it as comprehensively or formally as I need to. This I will do now in the form of a diagram of the relationships posited in the hypothesis:

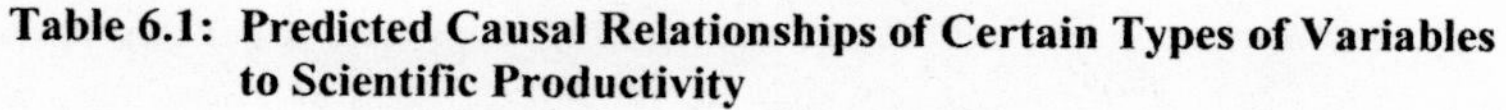

Table 6.1: Predicted Causal Relationships of Certain Types of Variables to Scientific Productivity

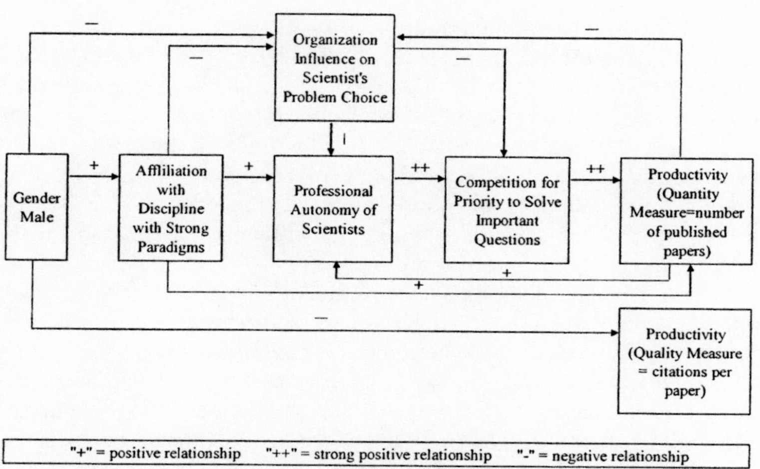

The reader may notice that the model restates the rationalist hypothesis of problem choice by scientists as a model of *productivity* of scientists." (*supra*). Consistent with the views of Valian, the model assigns *socialization* an important role as an exogenous variable. Here, however, discipline is introduced as an important mediating explanatory variable in its own right. Furthermore, organizational influence is made conditional on discipline which deviates from the position taken by Valian. The model suggested here also is different from Cole and Singer's stated position since, unlike Cole and Singer, it argues that socialization mostly is important for its effect on decisions as to which discipline to enter rather than on decisions subsequent to entry into the discipline. However, I believe that it is true to the spirit of Cole and Singer's views, which I regard as an eloquent argument that socialization is a crucial contributor to subsequent scholarly productivity.

The Impact of Gender on Productivity

I now would like to discuss the components of my hypothesis. The first relationship in the model I want to address is the hypothesized influence of gender on productivity (as measured by citations per paper). Before I discuss my reasons for this decision, I want to reiterate that gender is not biology but culture. Note also that I see men as socialized to care less about quality than women as indicated by the direct negative relationship between gender and productivity measured as citations per paper.

The demonstrated importance of gender differences in socialization in influencing productivity must be acknowledged. Cole and Singer (1991) and Valian (1998) both assign socialization important roles in explaining productivity differences between men and women. Given the support of prominent scholars such as these for an important role for socialization, I retain a *direct* influence of gender on productivity even though I am not intellectually comfortable with the idea that something in the distant past could exert a continuing influence on behavior in the present among scientists.

It is not just the support of prominent scholars, however, that lies behind my decision. Equally important, my stating a direct relationship is partly as an admission that something as yet unknown and unsuspected is influencing productivity. I do not have any plausible explanation besides socialization differences for why I found in Chapter Five evidence of differences in research styles of men and women scientists. For instance, Table 5.9 showed that both in their current and in their past research men were more likely than women to acknowledge trying to beat someone else for priority.[1] Furthermore, Table 5.19 showed that in selecting problems for study men were more likely than women to have taken into account access to special equipment or supplies. On the other hand, men were less likely than women to consider whether renowned researchers were urging study of the problem when selecting problems for study. Also, men cared less than women about hewing to deadlines for study completion than men did. The above findings which, I argue below, suggest possible reasons why women may score lower on productivity when only publication rates are considered, may be the result of socialization differences. And if socialization differences were indeed responsible for these differences, they certainly suggest an *indirect* role for socialization on productivity through socialization's influence on problem choice criteria, and research styles, and perhaps other as yet unidentified mediating variables.

The impact of Gender on Choice of Disciplinary Type

Whatever the reason, women scientists are found in fields with weak paradigms far more often than fields with strong paradigms. I do not see women as inherently less interested in mathematics, physics, chemistry, and engineering, for example, than men. However, historically, proportionally far fewer women have gone into these fields. That is changing and perhaps in the future women will enter these fields in numbers comparable to men. But a considerable literature (e.g., Zuckerman, Cole, and Bruer eds. [*The Outer Circle*]) has documented the hurdles women had to overcome in becoming scientists and, especially, engineers. Physics and chemistry and even may be more welcoming now but the small percentage of women professors in these fields, especially in the senior professorial ranks of leading research universities, testifies that women have not received much encouragement to pursue advanced degrees in those disciplines until very recently.

The Impact of Disciplinary Type on Productivity

The second relationship I call attention to is the direct and indirect influence of discipline type on productivity. I maintain that disciplinary type is essential to understanding productivity of scientists. There is ample evidence that discipline type is essential to understanding scientific productivity. Hargens (1975) and recently, Dietz (2004) showed that chemists produced research papers at a high rate-- a point that the latter further amplified. Neither of these researchers offered any *theoretical* explanation for their findings. This is surprising because the effect of discipline on productivity is considerably stronger in Dietz's data than is the effect of organizational variables. That, in itself, should justify the attention of anyone seriously interested in modeling productivity.

However, there may be an added dividend in determining the importance of discipline in productivity: it may prove helpful in assessing the role of gender in productivity. Cole and Singer's well-known effort to explain gender differences in productivity tried to assess the importance of gender differences by controlling in its data for disciplinary differences. This was an unwise decision for both theoretical and methodological reasons:

From a theoretical standpoint, deciding what field to major in is an earlier decision of future scientists—and perhaps a more important one from the standpoint of future research productivity-- than decisions such as choice of dissertation sponsor that Cole and Singer (1992;originally 1991) used to illustrate their argument.

From a methodological standpoint, Cole and Singer's attempt to assess the importance of gender differences by controlling for discipline and specialty was procedurally faulty as shown below. With reference to the theoretical reason, Cole and Singer pointed to "kick reaction pairs such as choice of dissertation sponsor, first academic assignment and tenure decisions. Undoubtedly, these are important to the aspiring scholar. However, we can legitimately ask, Are they demonstrably more significant in explaining a scholar's productivity than his or her initial choice of field? Given the considerable differences in productivity between fields reported by Dietz , it seems doubtful that choices made after the student scholars have entered their chosen fields have as much influence individually (and, perhaps, even combined) as the initial decision to enter one discipline rather than another. This is a question that needs to be answered with careful empirical research.

In regard to the methodological point, Cole and Singer made a questionable judgment in favor of trying to control the effects of possibly important variables (e.g. discipline) by *matching* samples. However, their matching procedure was faulty. For example, they conceded that they could not always match by *specialty* within disciplines. (In fn9, they report, "Where possible, men and women were matched by specialty at the time of receiving their degree)." However, even if they had been able to address this particular problem more adequately, they still could not be certain that they had obtained unbiased estimates of gender's influence on productivity since *current* discipline, as opposed to discipline in which degree was obtained, was not controlled.

In this chapter, I will try to assess the importance of disciplinary differences and specialty differences in order to understand gender's true effect on productivity. It is the position of this Chapter that researchers in fields characterized by "strong paradigms" will have greater research output than will researchers not so situated. (On paradigms, especially "dominant paradigms," and their role in scientific development, see Kuhn [1970; originally 1962]).

The Impact of Disciplinary Type on Professional Autonomy

Professional autonomy refers to freedom from non disciplinarian control of the problem finding process. There is never absolute freedom from such control. However, there clearly are degrees of such autonomy. The variable is measured several ways as was indicated in Chapter Five: e.g. the percent of respondents in a discipline who claimed to have freedom to pick their problems; whether a sponsor came and told the researcher what to study and even an index of such measures.

Disciplinary type indirectly affects productivity through its influence on professional autonomy of scientists. By professional autonomy I mean freedom from interference from others outside the discipline in such activities as conceiving, planning, and carrying out research in the field. There is no absolute professional autonomy but there are gradations in level of relative autonomy by gender and discipline. Men, I believe, enjoy somewhat more relative autonomy than women regardless of discipline. Disciplinarians in fields such as physics enjoy higher professional autonomy while disciplinarians in social sciences such as anthropology and sociology enjoy relatively less regardless of gender.

While I consider this view to be consistent with the intent of the rationalist perspective, it does mark a departure from the original formulation of the rationalist hypothesis (see Zuckerman (1978). Originally. the rationalist hypothesis implicitly assumed that all scientists enjoyed professional autonomy. The realism of this assumption has been challenged by numerous scholars (see e.g. Shenhav [1975]; Zeldenrust [1990] and I share their reservations (See Chapters One and Five, *supra* and the references cited therein).

The Impact of Professional Autonomy on Competition for Priority

Competition for priority is measured as an aggregated variable of answers to questions regarding whether a desire to beat someone in solving a particular problem was a criterion for doing the study. The higher the percent of such responses within a discipline or specialty the greater or more intense the competition for priority.

The relationship between professional autonomy and competition for priority is implicit in Zuckerman's (1978) argument that scientists free to choose problems will select the best problems they can based on their theory and methodological commitments thus giving rise to "intense" competition for priority as scientists rush to solve the most important problems in their discipline. Zuckerman made such a persuasive case for the impact of professional autonomy on competition for priority that even many critics of the rationalist hypothesis (e.g. Mulkey [1991]) accept this relationship.

The significance of Zuckerman's point, unfortunately, was overlooked by many scholars who focused on the dubious assumption of professional autonomy in her original formulation of the hypothesis. For example, earlier (see Chapter Five) I too criticized Zuckerman's formulation in part because her hypothesis predicts far more competition for priority than I found in my data. Cole and Singer (1992:284) attempted to improve on Zuckerman's hypothesis by suggesting that the primary goals of "priority for scientific discovery and accompanying peer recognition (Merton 1957); and success in the competition for resources to pursue research at a high level" were mediated by differences in gender socialization. This is an ingenious suggestion and probably has some explanatory value. However, it is necessary to take into account the type of discipline. And it is important to remember that professional autonomy is never absolute-- a point Talcott Parsons (1951:343) made over fifty years ago.

Parsons' point is worth repeating here. According to Parsons, the scientist occupies a "professional role" which is filled by people with "specialized knowledge" that "only with difficulty if at all [is] accessible to the untrained layman." Therefore, the scientist therefore must be given some freedom from external control (i.e. from persons others than his professional peers). Parsons later restated his point (1960:62-63) somewhat differently.

Regrettably, Parsons' astute observation has too often been ignored by later writers, perhaps because he framed it as sage advice rather than as an empirical finding of what actually occurs. As developed originally by Zuckerman (1978), for example, the rationalist hypothesis did not incorporate Parsons' insight. Both Zuckerman (1978) and Cole and Singer (1992), for example, assume that scientists are autonomous in their research in the sense of being answerable only to their professional peers.

In this Chapter as the diagram shows I will try to rectify the errors that resulted from ignoring Parsons' point. I argue that affiliation with a discipline with strong paradigms reduces the influence of employing organizations (especially of non- disciplinarians in employing organizations) on professional autonomy and willingness to engage in competition for priority. However, even if the influence of the organization is attenuated by these two factors, the organization still exerts some influence on willingness to engage in competition for priority and productivity (measured as a simple quantity).

It is worth mentioning that Parsons did not consider the issue of whether scientists from different disciplines could have relatively more or less autonomy depending on the strength of the disciplines' paradigms. But this is exactly what is implied by the model proposed here: different disciplines can be arrayed in a continuum based on the degree to which their paradigms are considered strong or weak. For instance, mathematics and heavily mathematical fields such as physics, chemistry, some branches of engineering and molecular biology qualify as fields where specialists will enjoy greater autonomy; sociology, many other social and behavioral sciences and medicine will enjoy somewhat lesser autonomy because non-specialists can more readily learn (or believe they already know quite a large amount) about the subject.

The important point in this discussion is that whenever there is less professional autonomy there will also be less competition for priority among scientists. Disciplines, or specialties within disciplines, whose practitioners enjoy relatively less autonomy will generally include not only those with weaker paradigms. They will also generally have a high proportion of women scholars (e.g. social work research, nursing research). Because many fields with weak paradigms also have high proportions of women scholars it is difficult to separate out the proportion of the productivity that is gender related versus the proportion that is related to discipline. However, I propose that gender and discipline type both have an influence on productivity.

The Impact of Competition for Priority on Productivity

Just as the relationship of professional autonomy with willingness to engage in competition for priority has been accepted since first proposed by Zuckerman, the relationship of willingness to engage in competition for priority with productivity has also generally been accepted. It is important to add, however, that this relationship occurs in the context of an organizational environment. I maintain that the organizational influence on both competition for priority and professional autonomy, especially among scientists working in fields with weak paradigms, can indirectly reduce productivity. However, it is an open question how much the organizational environment attenuates the productivity of the researchers. The extended discussion of organization effects on productivity in the literature (e.g. Feldt[1986], MIT [1999]) and the discussion of problem choice in Chapter Five makes clear that organization impact may be serious enough to reduce productivity measurably, especially among women scholars.

The Impact of Other Predictors on Productivity

Thus far, except for substituting productivity for problem choice, I have modified the rationalist hypothesis in only one major way: I have introduced the concept of disciplinary type and argued that it would make a measurable difference both directly and indirectly on productivity. However, the rationalist hypothesis, in my opinion, requires more modifications to fit the evidence. For example, I argue that gender has both a direct and indirect effect on productivity. I have already addressed the indirect impact of gender on productivity through its effect on type of discipline that scientists enter. It is important to say here that gender also has some direct influence. Quite simply, men scientists produce more papers than women regardless of discipline. This is only part of the story, though, as the diagram shows. Women scientists produce better research papers on average than men do as evidenced by more citations per paper. (see Valian [1998]).

Personal problem choice preferences and research styles are a potentially important factor in productivity that prior research has not even suggested. However, as was evident in the findings in Chapter Five of the study, personal problem choice preferences and research styles could influence how quickly problems are identified for research and how quickly the studies are performed once a problem choice has been made. (See Table 5.19 and the accompanying discussion, pp.202-204).

The Impact of Productivity on Professional Autonomy

I maintain that productivity has a delayed effect on professional autonomy. There are some scientists in disciplines who enjoy high autonomy even when most practitioners in the field enjoy more limited professional autonomy. I believe that the high productivity of scientists can increase their professional autonomy. This point acknowledges the truth in the old adage advising people not to slaughter the goose that lays golden eggs.

Evidence for the Modified Rationalist Hypothesis

The relationships proposed here, even if plausible, cannot all be evaluated with the data on hand. However, I will present data pertinent to some of the key relationships below. Before I do that, however, I want to say that the data for this study consisted of survey data from 107 scientists from various disciplines gathered to test the rationalist hypothesis of Zuckerman against other models of *problem choice in science*, not productivity. More details about the data can be found in Chapter Three *supra*.

A. Autonomy and Discipline Type.

Because a central assumption of the rationalist hypothesis is that scientists are professionally autonomous in problem choice, several indicators of autonomy were included in the questionnaire (also used as an interview protocol). In Chapter Five, though I showed that men and women differed in their professional autonomy, I did not look at disciplinary differences. When these autonomy differences between genders were controlled by discipline type, there were no statistically significant differences between men and women in fields with strong paradigms. However, as Table 6.2 shows (see below), in fields other than those with strong paradigms there were large statistically significant differences in the professional autonomy of men (91%) and women scientists (61%) in the data set.

Table 6.2: Freedom to Choose Problems by Gender of Scientists (Disciplines with Weak Paradigms Only)

	Gender of Scientists			
Free to Choose	Men		Women	
	Percent	Number	Percent	Number
Yes	91%	41	61%	20
Total		45		33
	$X^2 = 10.39$ P = .001 Mantel - Haenszel = 10.26 P = .001 N = 78			

That gender is a predictor of professional autonomy in disciplines with weak paradigms but not in disciplines with strong paradigms may be an essential point—and one that the debate about gender differences has so far failed to take into account. If corroborated in other studies, it would have the salutary effect of (a) shifting attention among the scholars of productivity to understanding why women with a scientific bent choose to major in particular fields and avoid others when they attend college. And (b) it would begin to test the Cole and Singer (1992) hypothesis of "limited differences" in socialization although Cole and Singer overlooked this possibly crucial decision in discussing influences on future research productivity.

B. Competition and Productivity

Cole and Singer (1992:305) ascribe an important role to competition in science, at one point characterizing their theory of "limited differences" as "in many respects ... a theory about the response of the community of scientists to an unstated driving force: competition." There is competition for scarce resources, the kind familiar to economists, and competition for priority. The latter, Cole and Singer assert, occurs "for some problems, especially in the upper echelons of any scientific discipline, with the competing parties having nearly complete knowledge of what their competitors are doing"(p. 306). Cole and Singer then make the interesting observation that

> ... the intense transfer of information" via frequent conferences, private laboratory visits, and even telephone conversations between competitors and/or their close collaborators plays a major role in structuring research

> agendas and in regulating the duration of time between experiment completion, manuscript completion, and publication."

Cole and Singer's contention lucidly states the rationalist assumption of "perfect information" underlying the competition for priority—an assumption justified in the rationalist view because (of a further assumption that) research problems are "previously identified" (Zuckerman [1978:]) or "shared" (Gieryn [1978]). Those assumptions, however, have been criticized both here and elsewhere (see Zeldenrust [1990]). For example, in Chapter Five, I argued that the twin assumptions of "shared problems" and perfect information are more justified in fields with strong paradigms (e.g. high energy physics) than otherwise.

I tested my hypothesis about the relationship of "shared problems" and perfect information on problem significance by (1) taking as an indicator of "shared problems" scientists' response that they compared problems before choosing the one they decided to research and (2) taking as an indicator of "strong paradigms" scientists' response that they had derived their research problem from theory. Proportionally more scientists among those claiming their problems were theoretically derived compared problems before selecting one to investigate than among scientists who did not make this claim. (Table 5.5. p. 172).

I also explored the role that competition played in influencing problem choices and productivity as Cole and Singer suggested. The evidence did not indicate that competition for priority was a predictor but neither did it conclusively show that competition played no role. There could be several explanations for these ambiguous results, including insufficient number of cases. On the other hand, it is certainly true that in many fields, scientists do not feel that they are in competition with anyone else.Mulkay found this in his study of radio astronomers in the United Kingdom. (Mulkay, 1991:originally 1974) Furthermore, it is possible to find a specialty so unusual that no one else is interested in it, e.g. the study of certain so-called "orphan diseases." In these "quiet fields," scientists probably have little or no competition to solve the same problems they are investigating.

In other fields, competition for resources among the scientists is quite pronounced and competition for priority becomes the focus for that competition for resources. For example, in the race to discover a "cure" for dementias, especially Alzheimer's syndrome, scientists are working intensively to find genes whose mutations will help them better understand the dynamics of brain deterioration. They and/or other scientists with whom they are associated are also racing to develop potentially effective chemotherapeutic agents that compensate for the effects of the gene mutations thought to be behind the development of these dementias. Thus, it is clear that competition for priority is associated with competition for resources even though intellectually they can be considered separate kinds of competition.

One intriguing finding emerging from the exploration of competition was that men are more likely than women to acknowledge competing for priority with others in their research. A number of reasons may account for these differences. For example, possibly, men gravitate to fields where competition is keener—or maybe women simply are less candid (because of the way they are socialized?) about their competitive natures. Table 6.3 below shows that among scientists in fields other than those with strong paradigms, men were considerably more likely than women to acknowledge that a desire to beat out a rival played some role in their problem choice. This was true in both reporting about the criteria they used in selecting their most recent project and in reporting on past research in which they were involved. While the gender differences are notable, perhaps the most important finding is that only a minority of respondents of either gender claimed that competition was a consideration in their problem selection (and possibly therefore in their procedure for turning research into published articles etc.).

Table 6.3A: Wanted to Beat Rival in Most Recent Project by Gender (Disciplines with Weak Paradigms Only)

	Gender			
Wanted to Beat a Rival in Most Recent Project	Men		Women	
	Percent	Number	Percent	Number
Yes	32%	14	6%	2
Total		44		33
	Fisher's Exact Test (1-sided) P= .001 N = 77			

Table 6.3B: Wanted to Beat a Rival in Past by Gender (Disciplines with Weak Paradigms Only)

	Gender of Scientists			
Wanted to Beat a Rival in Past	Men		Women	
	Percent	Number	Percent	Number
Yes	47%	21	24%	8

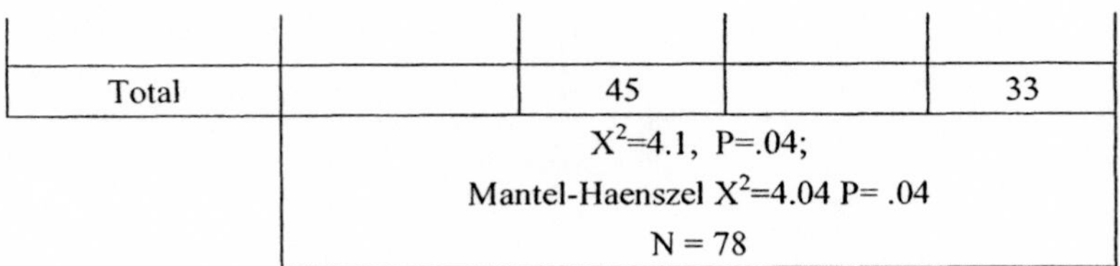

Total		45		33
	X^2=4.1, P=.04; Mantel-Haenszel X^2=4.04 P= .04 N = 78			

Considering that in no disciplines was the desire to beat out a rival a factor in problem choices for most respondents it seems that competition only played a minor role in problem choice and productivity. What might explain the higher productivity of some disciplines than others? One clue might be the "intellectual climate" of the discipline. In response to a question regarding their discipline's "intellectual climate," a heavy majority (67%) of those in disciplines with strong paradigms claimed that the "intellectual climate" was "turbulent" whereas only a minority (40%) of those in other disciplines felt this way. (See Table 6.4 below).

Table 6.4: Intellectual Climate by Type of Discipline

	Type of Discipline			
	Strong Paradigms		Weaker Paradigms	
Intellectual Climate	Percent	Number	Percent	Number
Turbulent Climate	67%	18	40%	29
Total		27		72
	X^2=8.76 P=.01 Mantel-Haenszel X^2=6.15, P=.011 N=99			

These results seem at first glance to be counter intuitive. My first inclination would have been to assume that disciplinarians in the social sciences, heavily represented in the small data set used here, would find their disciplines "turbulent". This is because many disciplines in the social sciences are characterized by weak paradigms; most topics of wide interest in the social sciences are characterized by vigorous debates among scholars from different camps. (There are even debates about whether disciplines such as sociology are sciences at all). This chronic situation has given rise to the feeling that a fundamental problem of the social sciences is that they suffer from a "plethora of theories and a sparseness of data." Despite this, far more respondents in chemistry and physics, disciplines with strong paradigms, regard their disciplines as having a turbulent climate than is the case with respondents in the disciplines with weak paradigms.

The reason for this finding, I suggest, is that the disciplines with strong paradigms may be on the verge of embracing new paradigms-- or at least highly modified versions of paradigms that have been generally accepted for a long time, (decades in some cases). There is a consensus building that the old paradigms are not satisfactory in important ways but there is no agreement yet about what if anything is better. Numerous alternative models are vying for the honor of being the new paradigms; some of these mere modifications of the old paradigms while others such as "superstring theory" are truly novel. Among scientists raised on Einstein's relativity theory or Glashow's subatomic model based on "quarks," the current state of affairs in physics must indeed seem turbulent. It is only from the vantage point of a discipline such as sociology that is waiting for its Newton (or even its Darwin, as William J. Goode once acerbically noted), that physics seems orderly.

The foregoing discussion indicates that the rationalist hypothesis oversimplifies the reason for the competitive environment of science and for the high productivity of scientists. Contrary to what the rationalists have argued, it is not the existence of strong paradigms per se that leads to intense competition for priority. Rather, it seems that the many conflicting hypotheses suggested by competing models vying to become the new paradigms are accounting for the intensely competitive and turbulent environment of the disciplines with "strong paradigms." Secondly, in the disciplines without strong paradigms, perhaps the proportion of disciplinarians of the male gender may influence the competitive environment. Men are more likely to be competitive (or at least to admit to being competitive). Consequently, in a discipline (or specialty) that is preponderantly male, disciplinarians may be more prone to engage in competition for priority than in one with a higher proportion of women scholars.

Does the competitive nature of the discipline bear on productivity, as suggested by Cole and Singer (1992)? I believe that it does but mostly because of the conditions giving rise to that competitive environment. The existence of a number of alternative theories each with the potential to explain a lot of phenomena and in conflict with one another is a powerful impetus to research. And, of course, this impetus to research also gives rise to an intensely competitive environment as scientists committed to different models produce research papers, critique the work of their rivals, and so on. Furthermore, a preponderance of men researchers in the specialty or discipline may also have a bearing on the competitive atmosphere within the discipline. And, finally, the possibility of large financial gains from the fruits of research may by itself give rise to a competitive atmosphere and high productivity. The person(s) who finds a cheap high temperature superconductor or discovers a low cost cure for AIDS etc. can earn millions of dollars or even more from the discovery. Talented scientists will be drawn to such remunerative work just as other entrepreneurs are drawn to projects likely to have high payoffs relative to the assessed risk.

C. Organization Influence in Productivity

While I have argued that organizational influence on problem choice could be either positive or negative, I share the rationalist view that as regards its influence on productivity it is invariably negative—albeit not strongly so. In disciplines with strong paradigms, for example, individual scientists are unlikely nowadays to make the big discoveries. Massive resources are needed to conduct the type of research in high energy physics that is required to test the new theoretical models. Inevitably, bureaucracies must be involved in the research and that slows the work simply because of the multiple parties to be coordinated. In fields with weak paradigms, the role of bureaucracies also is probably at least slightly negative. In these fields, to the extent massive resources are needed, the same reasoning applies. However, in fields with weak paradigms, there is the added complication that non-specialists have a bigger say in what gets studied than in fields with strong paradigms. Put another way, scientists' probing for demand for research on a topic and grantors searching for problems to address their concerns (see Zeldenrust [1990]) is a more pronounced feature of the landscape in fields with weak paradigms and this also negatively influences scientist productivity.

The reader may recall Valian 's tart comment (1998) that the culture of elite institutions does not just encourage scholars to be more productive it forces them to be, Here, it is pertinent to ask the correlated question, Is it also true that weaker (*i.e.*, non-elite) institutions depress their scientist employees' desire to produce and publish? After all, it seems reasonable to suppose that graduate school training should ignite the students' desire to contribute in a scholarly way by doing research and publishing it.

However, as indicated in Chapter Five regardless of whether the scientists were at an elite institution women scientists experienced more interference in their research than men (see Table 5.20). This could lead to women's productivity being adversely affected . Was the likelihood of interference in research work greater at weaker or non-elite institutions than at elite institutions? If (a) women were concentrated at non elite institutions while men were concentrated at elite institutions, and (b) the elite institutions interfered less in scientist work, then it follows that the productivity gap indeed is related to organizational culture as Valian predicts.

Suppose that women experienced more interference than men regardless of the prestige of the institution. Presumably, this would depress women's productivity compared to men's. This would suggest that an informal norm in (some or all of the) scientific disciplines rather than organizational culture was the culprit.

These hypotheses can be compactly presented in the form of a typology (see below) in which the cells represent expected scientific productivity levels. Adding a third controlling dimension of gender would create an eight-fold table (or two four-fold tables, one for each gender), a point we will pursue further below.

Table 6.5 Predicted Levels of Scientific Productivity Given Variations in Institutional Prestige and Organizational Culture

	Institutional Prestige	
Organization Culture	Elite Institution	Non-elite Institution
Non-interfering	(1) High	(2) Med-High
Interfering	(3) Medium	(4) Low

Table 6.5 merely asserts that productivity monotonically decreases from the first cell at the upper left (elite institution/non interfering culture) through the fourth cell at the lower right (non-elite institution/ interfering culture).

I hypothesize that interfering organization cultures depress productivity. If women are concentrated in institutions that are more interfering, or if all institutions interfere in women's research more than in men's, I hypothesize that women's productivity would be adversely impacted compared to men's. I also hypothesize, however, that when organization prestige is higher, productivity of women is higher than it would be at a low prestige institution.

I further believe that an interaction can occur between either or both of the two organization dimensions of organization culture and institutional prestige on one hand and gender (and/or discipline) on the other. Several possible ways for women's productivity to be reduced compared to men's can arise out of the interactions:

(a) the productivity of women scientists is reduced because they experience interference in their research across all research settings. In other words, "non-interfering cultures do not exist for women.

(b) A further possibility is that institutional prestige is irrelevant to the productivity of scholars.

(c) The productivity of women scientists is reduced because they are concentrated in low prestige institution whereas men are concentrated in high prestige institution.

(d) A further possibility is that non-interfering cultures do not exist for anyone.

Note that Cole and Singer (1991) reject the possibility of differential interference by gender. Therefore, it is certainly important for the viability of their hypothesis that the data not support possibilities (a) or (c). I then further observed that regardless of whether the scientists were at an elite institution women scientists experienced more interference in their research than men (see Table 5.20 *supra*). This interference could lead to women's productivity being adversely affected.

According to this logic, the productivity of women scientists is reduced because (a) they experience interference greater than men and (b) they are concentrated in low prestige institution. It is not necessary, that the organization culture be uniform across the entire university or research center. Some organizations, to use an old term, are elastic enough to have quite varied organization cultures under the same roof. This is possible because the components are so widely distributed (e.g. the Roman Catholic Church) that conflicts between parts can be kept from tearing the organization apart. Thus, depending on the department within the university, for example, the organization culture could be interfering or not within the same overall organization. The organization culture, also could be experienced differently by different genders within the same organization.

It is now time to look at the evidence on the impact of organization traits on productivity. No claim is made here that this is an exhaustive review of the organization's influence on scientist productivity. However, I believe the evidence on the point is interesting.

Table 6.6 shows that when organizations rather than the principal investigators picked the problems the researchers' productivity was lower than when the scientists picked the problems. The Table also shows that lower scientist productivity was associated with the organization wanting the political sensitivity of the problems considered in problem choices. While these findings are certainly consistent with a view that organization interference reduces productivity, we must be careful not to read too much into them. For example, these findings are also consistent with a view that the scientists' lower productivity might be causing the organization to be interfering. Thus, the causation could go in either direction or even be two-way.

Table 6.6A: Productivity by Organization Picks Problems

	Organization Picks Problems			
Productivity	Yes		No	
	Percent	Number	Percent	Number
Lowest	36%	18	17%	4
Low	32%	16	12%	3
Medium	16%	8	38%	9
High	16%	8	33%	8
	$X^2 = 9.96$ P = .02 Mantel - Haenszel = 7.24 P = .007 N = 74			

Table 6.6B: Productivity by Organization Considers Political Sensitivity

	Consider Political Sensitivity			
Productivity	Yes		No	
	Percent	Number	Percent	Number
Lowest	38%	13	16%	7
Low	29%	10	23%	10
Medium	21%	7	27%	12
High	12%	4	34%	15
	$X^2 = 8.34$ P = .04 Mantel - Haenszel = 8.22 P = .004 N = 78			

Problem Choice Criteria and Research Styles: Their Influence on Productivity

In Chapter 5, I presented data that demonstrated gender differences in research criteria used to select problems for study (see Table 5.19). Some or all of them might influence research productivity at least when it is measured in purely quantitative terms. For example, other things being equal, it takes longer to assess a problem's feasibility and worth if you must be concerned with:

- Whether respected colleagues are urging work on it or;
- Whether the problem is related to a public controversy or;
- Whether the research project can be completed within applicable time constraints.

It is also germane to point out that, regardless of discipline, men respondents were more likely to report their willingness to take advantage of available special equipment or supplies for their research than women were. Men's greater willingness to ignore convention in this regard may also help account for their greater productivity. According to Table 5.19 men were twice as likely (50%) as women (25%) to weigh whether sophisticated equipment was available for their research in deciding what sorts of problems they wanted to study.

Beyond these findings in a small data set, there was no prior literature suggesting any connection between productivity on one hand and problem choice criteria and research styles on the other. However, productivity is a subject of intense interest in sociology for which no satisfactory model has yet been developed. Despite limitations of the data set, it seemed worthwhile to find out what could be learned about the influences on productivity in these data. At the very least, it could assist in planning a large scale study to resolve the productivity puzzle.

Therefore, I also looked at the personal research styles and problem choice preferences of scientists to see if these contributed to productivity. The findings in this area are more clear cut in their theoretical meaning than those regarding organization traits' relationship to productivity.

Table 6.7a shows that scientists' taking into account the political and/or social significance of a problem was associated with lower productivity among my respondents.

In addition, wanting to work on a next project with other colleagues from whom the scientist could learn was associated with lower productivity. (Table 6.7b) Seventy-eight percent of scientists stating that their problem choice in their next project was influenced by the possibility of working with someone with whom they were eager to work were in the two lowest classes compared to only 40% not claiming a desire to work with someone on their next project.

Making problem choices urged by renowned colleagues was likewise associated with lower productivity. (Table 6.7c). On the other hand, a strong desire to contribute a technical paper in a new dynamic area was associated with higher productivity.(Table 6.7d).

Table 6.7A: Productivity by Socially Significant Issues

	Socially Significant Issues			
Productivity	Yes		No	
	Percent	Number	Percent	Number
Lowest	35%	21	20%	9
Low	27%	16	18%	8
Medium	21%	13	27%	12
High	17%	10	35%	16
	$X^2 = 6.89$ P = .076 Mantel - Haenszel = 6.47 P = .011 N = 105			

Table 6.7B: Productivity by Next: Work With Colleagues Eager

	Next: Work With Colleagues Eager			
Productivity	Yes		No	
	Percent	Number	Percent	Number
Lowest	49%	17	22%	10
Low	29%	10	18%	8
Medium	11%	4	30%	13
High	11%	4	30%	13
	$X^2 = 10.68$ P = .014 Mantel - Haenszel = 9.34 P = .002 N = 79			

Table 6.7C: Productivity by Important by Renowned Researchers

Productivity	Important by Renowned Researchers			
	Yes		No	
	Percent	Number	Percent	Number
Lowest	30%	15	27%	15
Low	30%	15	16%	9
Medium	24%	12	24%	13
High	16%	8	33%	18
	$X^2 = 5.16$ P = .16			
	Mantel - Haenszel = 2.53 P = .11			
	N = 105			

Table 6.7D: Productivity by Publication In New Area

Productivity	Publication In New Area			
	Disagree		Agree	
	Percent	Number	Percent	Number
Lowest	27%	19	34%	13
Low	19%	13	29%	11
Medium	24%	17	21%	8
High	29%	20	16%	6
	$X^2 = 3.37$ P = .34			
	Mantel - Haenszel = 2.46 P = .12			
	N = 107			

These simple bi-variate relationships convinced me that productivity might indeed be influenced by organizational variables and personal problem choice preferences and research styles of scientists. However, it was necessary to scrutinize the relationships found in a more searching way. Multiple regression analyses seemed an obvious choice for this purpose. A multiple regression model would have numerous advantages as (1) identify significant predictors; (2) assess the strength of a covariate and (3) assess the quality of fit of a model.

These were all desirable features since I wanted to know if there were good reasons from a statistical standpoint to believe that the gender gap in productivity was related to:

(1) individual scientist problem choice preferences and/or;

(2) organizational variables.

Both individual preferences and organization variables had been suggested by prior theory (e.g., Valian [1998] and/or survey findings of the senior author's study of problem choice. If the multiple regression analysis did show that either of those types of variables were relevant, that would represent progress in solving the problem of why there is a gender gap in productivity – an "old puzzle," as Xie and Shauman (1998) describe it, for which no satisfactory solution has yet been found.

The reader should remember the limitations of the analysis as she reviews results of this multiple regression analysis presented below.[2] A multiple regression analysis is only as good as the data on which it is based. In this analysis, as in cross tabulations presented above, lack of sufficient cases precludes treating the results of the analyses done here as a comprehensive model that correctly estimates the impact of all relevant indicators. Another possible problem is that these data were not collected for the purpose to which I am now applying them. I have tried to bolster the validity of the findings where I could. Sometimes, for example, I was able to compare results I obtained with those of other studies (e.g. Dietz [2004]) and the similarity of the results suggested that I had probably identified important predictors (though others remain to be found).

THE MULTI-VARIATE ANALYSIS OF PRODUCTIVITY

1. The Variables

The outcome variable, productivity ("PROD") was defined to be the number of articles plus *three* times the number of books. PROD was an integer outcome whose values depended on the number of years of experience. The predictors were variables that assessed the motivation the respondent has for producing books or articles.

Below is a list of "yes, no" variables tested for their relationship with productivity. Variables with asterisks (*) to the left of them are those that appeared in the selected model.

- **Prefer to do what others in field are doing.** This was a multi-category variable that was recoded in several dichotomous versions:

Prefer what others-A: 1 = always; 0 = all others
Prefer what others-B: 1 = most of the time/usually; 0 = all other responses.

Prefer what others-C: 1 = sometimes; 0 = all other responses
Prefer what others-D: 1 = rarely/never; 0 = all other responses

➢ **Prefer Problems No One Else is Likely to be Working on:** This multi-category variable was recoded as follows:

Prefer Problems No One Else-A: 1,2 = 1 (always/most of the time); all else = 0
***Prefer Problems No One Else-B:** 3,4 = 1 (usually/sometimes); all else = 0
Prefer Problems No One Else-C: 5,6 = 1 (rarely/never); all else = 0

➢ **Desire to Contribute Something of National Value:** This multi-category variable was recoded as follows:

Desire to Contribute-A: 1 = disagree strongly, disagree or neutral; 0 = all others
Desire to Contribute-B: 1 = agree mildly through moderately; 0 = all others
Desire to Contribute-C : 1 = strongly agree; 0 = all others

➢ **Desire to Publish a Technical Paper in a New Dynamic Area:** This multi-category variable was recoded as follows:

***Desire to Publish in New-A:** 1 = disagree strongly, disagree or neutral; 0 = all others
***Desire to Publish in New-B:** 1 = mild to moderate agreement; 0 = all others
***Desire to Publish in New-C:** 1 = strong agreement; 0 = all others

➢ **Could associate with technically superior colleagues (MDR)** Only one recode was performed on this multi-category variable

***Could associate with technically superior -A:** 1 = agree mildly to strongly; 0 = all else
N.B.: Cases coded "0" in the original variable were not included in the

analysis of regression.

➢ **I chose the MDR project because it gave me a chance to work with experts in the field).** This dichotomous variable was not recoded (except to make "2 = no" into "0 = no"").

➢ **Could study problem within applicable time constraints** This dichotomous variable was not recoded.

Finally, gender (SEX) was also entered in as a 2-value variable

SEX: 1 = male 0 = female

(N.B: values 7 = "unknown"; 8 = "no opinion"; and missing were <u>deleted</u> entirely from all dichotomous versions of each predictor variable).

2. The Choice of the Model

A Quasi likelihood model with scale parameter was chosen as the best model after deleting three observations from the data set. The model is:

Ln (rate (X)) = .4242 + .7739(Prefer Problems No One Else-B) + .6772(Desire to Publish in New-B) + .5386SEX - .6243(Could Associate with Technically Superior-A)

The values of the included variables were all described earlier. (see the * variables in the list). Below we discuss the reasons for our choice of this model.

The objective is to model the productivity with the predictors, and determine which predictors are influential. The productivity of a respondent depends on the number of articles plus three times the number of books divided by years of experience: YEARSEXP. The greater YEARSEXP is, the greater the variability there is for the productivity. Thus a normal model, typically with equal variance for each observation, is not appropriate: here the variance is potentially different for each observation. Poisson regression with the log link function models the log of the rate of productivity as a linear combination of the predictors: $ln(R(X)) = \sum \beta jXj + ln\ Nj$, where R(X) is the rate of productivity for the covariate pattern Xj and Nj is the years of experience of person j. Therefore, while the observed variable PROD is in fact the respondent's production, by including years of experience as an offset variable the parameter is the rate of production i.e. productivity. And consequently, productivity R(X) is a multiplicative model of the factors e^{BjXj}. This model is commonly known as the log linear model for Poisson outcomes.

There were other possible choices for a model with count data: negative binomial, quasi-likelihood, and a gamma model. All these productivity models were log linear and only differed in how the variance depended on the mean of the response variable, PROD. An examination of these models showed that there was a strong consensus across all of them that certain predictors were statistically significant. [I shall have more to say about this below].

The Poisson regression model was rejected because it fit poorly as evidenced by considerable extra Poisson variation dev/df=1832/82>22. Thus both the deviance test and the Pearson test would reject the adequacy of fit with a p- value less than .001. The quasi likelihood distribution defined by taking the variance of PROD to be equal to the mean of PROD times a scale factor, $V(PROD_1) = E(PROD_1)\text{Ø}$, was the next model in the hierarchy of variance models. Here, as in all other quasi likelihood models, *Ø* the scale parameter, was estimated by the Pearson statistic: $\hat{O}$ = (Pearson) / *df.* In this case $\hat{O} = 21.25$. The deviance and Pearson tests of the adequacy, necessarily, will not be a suitable basis for rejection.. An examination of residuals is required to assess the quality of fit.[3]

The examination of the standardized Pearson and deviance residuals showed that there were three observations that had large residuals. These residuals were influential on various outcomes in the analysis: strength of significance of the predictors (p-values), standard errors of the coefficients and the deviance. Only one of the three large residuals belonged to a very large value of PROD. Therefore, these residuals were not an effect due to extremely high productivity. They were either statistically unusual values for that covariate pattern or the result of predictors not found in the data. The quasi likelihood model with $V(PRODj) = E(PRODj)\text{Ø}$ was obtained in which the three residuals were deleted. The fit was better. The likelihood ratio test that the three observations, which were deleted, made a significant contribution yielded a highly significant result of $P \leq .001$. Since it could not be demonstrated that the observations considered for deletion belonged to an under represented population of respondents that have a very large number of publications it was decided to include these observations in models under consideration. Thus *there may be important co-variates which were not included* in the survey.

The next models considered were the negative binomial model and the Gamma model, both with logarithmic links. These log linear models for the rate of productivity assume that the variance of the outcome, PROD, depends on the mean squared. The same three observations have large residuals in the negative binomial and the gamma models, and both models indicated a lack of fit by displaying under variation: dev/df is too small.

Finally, the choice of predictors in all three models was examined. The Quasi likelihood Poisson with a scale parameter fit at least as well as any of the other models considered. Since the Quasi likelihood Poisson deleting the three observations with large residuals chose nearly the same predictors as all the models (Poisson, negative binomial, Gamma) including the outlier observations, the Quasi likelihood model with scale parameter (and without the three observations) was deemed the best model.[3]

Identification of Significant Predictors

The first objective of this study was to identify any significant influences on productivity from among organizational characteristics and scientist attitudes. Typically, the variables selected as predictors depend on the choice of predictor model. And as theorizing in social science is at an early stage there is sometimes a temptation to choose the model that agrees with one's own prejudices secure in the knowledge that probably no other model is clearly superior. What is interesting in the present case is that there was a consensus across all the models examined that certain variables were having an impact on PROD.

In this discussion, I omit results from analyzing the Poisson model with no scale parameter and from analyzing the Gamma model. In the former, the model exhibited too little variation and the result was that all predictors entered were highly significant. In the latter case, the model also exhibited under variation. This was due to the dependence of the Gamma model's variance on the mean being larger than it should have been.

I report here results of the predictors identified in (a) the quasi likelihood Poisson model with a scale parameter (minus the three outlier observations); the same model including the three outlier observations; and the negative binomial model. Across all three models, all the variables selected in the best model were always Wald and Likelihood ratio test significant at the .05 level and typically much less than .05.

Furthermore, across all three models there was agreement about the sign and the absolute value of the coefficients of the selected variables.

Because there were no interaction terms included in the model we selected, it is a straight forward computation to identify the effect of each of the predictors on the rate of productivity: for covariate pattern $X_0 = (X_{10}, X_{20}, X_{30}, X_{40})'$ the proportional change in the rate of productivity for variable Xj is $R(X_0, Xj_0=xj_0) - R(X_0, Xj_0=0) / R(X_0, Xj_0=0)$.

As indicated in the author's perspective (*supra.)* I was not intellectually pleased with any model that assigned socialization a direct role in productivity. I hoped that the totality of the variance of productivity (PROD) would be predicted by organization characteristics and personal research problem preferences of the respondents. I wanted a model showing socialization as at most a deep background variable having some influence on the personal research problem preferences of respondents but *not* directly on PROD.

If the initial hypothesis were correct that organization characteristics and scientist research preferences were the only direct predictors of productivity, I expected that gender – as the proxy for socialization – would not appear as a predictor in the best model. The influence of gender on PROD was investigated by first modeling it with the quasi likelihood Poisson model with scale parameter but without including gender (SEX) among the predictors.

Thus, PROD***i***, the predicted value of PROD for observation ***i*** is the expected value of PROD using the model without SEX***i*** ; and PROD***i*** – PRED***i*** is the additional productivity for person ***i*** after adjustment of the other predictors.

If gender were not a significant predictor for PROD, then the two samples of residuals PROD***i*** – PRED***i*** constructed for males only and for females only should resemble two samples from the same underlying distribution. A computer assisted comparison of these two samples with a standard statistical software package (SAS's univariate procedure, PROC UNIVARIATE), made it abundantly clear that the two samples were from different populations. Thus, in the component population, SEX = 1 (males), there was an extreme observation. After deleting this observation, the means appeared different: -8.8 for women and 2.63 for men, but the variances were large. This meant that a two-sample test would not be appropriate for testing the differences. A non-parametric test or resampling test seemed to be justified and the Satterthwaite test was used. The model did show SEX to be a significant predictor, but more work could reasonably be done on the association between PROD and SEX.

Once it was clear that gender (SEX) needed to be present in the best model, PROD was modeled using the same variables and SEX to obtain the model coefficients. I now address each of the predictor terms of the model.

The variable "Prefer problems no one else is likely to be working on—B," is difficult to interpret. This is because when the variable has a value of one it is a double negative: it expresses preference for problems others are working on by denying a desire to work on problems no one else is working on. Simply put, there is a 117% increase in the rate of productivity of persons who prefer working on problems others are also working on. That is, researcher productivity is enhanced when researchers are working on problems others also are interested in solving.

"Desire to Publish in New-B" measured the strength of agreement with the statement "I want to contribute to a research project leading to publication of a technical paper in a new dynamic area." Desire to Publish in New-B" = 1 means that the respondent strongly agrees with this statement. Those respondents who strongly agreed that they wanted to contribute in this fashion had a 96% increase in the rate of productivity over those who did not strongly agree.

SEX is the variable name given to gender in this analysis, the proxy for socialization. SEX = 1 means that the respondent is a man.

The effect of being a male respondent was to increase the rate of productivity by 55% in this study. Gender, possibly a surrogate variable for many other factors, was the most consistent predictor for productivity as it appeared in every model tried as a significant predictor. Thus in addition to the gender specific variable identified by Cole and Singer (1991) and Xie and Shauman (1998) there may well be other gender specific variables that are predictive of productivity after gender specific disadvantages have been address.

Respondents whose reasons for choosing MDR projects did not include the fact that participating in MDR allows association with technically superior colleagues had a 46% higher rate of productivity than those for whom this was a consideration.

Discussion

If research styles and problem choice preferences are important reasons for gender differences in productivity, does this mean that women should change their research styles and problem choice preferences to more closely resemble those of men, (or, more accurately those of the more productive men)? I do not think this is a correct inference from these findings even if subsequent studies confirm them. The findings here should be considered in the light of findings of several studies (see Valian [1998];also see Long[1992] Sonnert[1995];and Zuckerman[1987]) that women do higher quality studies than men as measured by citations per published paper.[4]

Women are following a more conventional strategy of research because they need to. Sulloway (1996) offers an intriguing reason that the second born (or later arrival) needs to differentiate itself from the first born (or first arrival) in order to gain investment in itself from the caregiver (or other financial supporter). A more prosaic reason for women's differences in research style and problem choice is simply that as latecomers to the academic world, especially in fields such as engineering and many of the exact sciences, women scientists are keenly aware that their work is regarded more skeptically than men's research. Women scientists understand that not only men in their discipline may be looking more critically at research by women scientists. Women colleagues will also be quick to condemn the low quality work of women scientists. This is because these colleagues are afraid that poor quality work by women will provide ammunition to those hostile to women in the discipline.

Therefore, these findings of differences in research styles and problem choice criteria between men and women should be seen not as a reason for women to change their research strategy. Rather, I believe that organizations may need to change how they evaluate women professionals' work. Comparing men and women on purely quantitative measures of productivity is unfair to women even if such purely quantitative measures do not at first glance appear to be discriminatory. (It is possible that comparing white men and anyone from other protected category groups that have not been represented in the scientific discipline in substantial numbers - at least until recently – is also unfair to these protected category groups for similar reasons). A fairer system of evaluating men's and women's scientific contributions would first consider (1) quality measures such as citations per paper and then (2) quantitative measures only *within* the gender group (or *within* the protected category and *within* the non-protected category).

Conclusions

The main contribution of this study of researcher productivity has been to demonstrate that scientists' personal research styles may be independent predictors of productivity, even in models including gender as a predictor of productivity. This finding may point the way towards solving what Cole and Zuckerman (1984) labeled the "productivity puzzle" of a gender gap in publication rates of men and women scientists. This gender gap, first noticed in the 1960s, continues at least in some disciplines to the present time (see Xie and Shauman[1998], causing them to refer to it as an "old puzzle."

Prior attempts to understand the productivity gap have generally emphasized socialization issues, e.g. the "limited differences" perspective of Cole and Singer (1991) or the "resource differentials" perspective of studies such as Feldt (1986) or MIT (1999). Previous studies that hinted at a role for organizational characteristics (e.g. Long[1992]; Xie and Akin [1994] and Xie and Shauman [1998]) in my opinion belong with the "resource differentials" perspective. Valian (1998) combines both the resource differentials perspective and the socialization perspective in her review of the gender gap issue. However, as Xie and Shauman (1998) observe, thus far an entirely satisfactory explanation has "remained elusive' (Xie and Shauman [1998:847]).

The present study examined the possible influence of selected organizational characteristics on productivity including "whether the organization wants the political sensitivity of problems considered before research is commenced" and "whether the organization superiors select problems that the researchers work on." The evidence of an effect were weak. Though in some simple cross-tabulations I found statistically significant differences in productivity I found no effect in the results of multiple regression analyses. Possibly, this was the result of having too little data. Furthermore, because there were insufficient data the study did not investigate the impact on productivity of the two variables (organizational prestige and organizational culture) that prior theorizing suggested were important determinants. At this point I have no position on whether organizational variables ultimately will be found to be direct influences on productivity.

I also sought in this study to determine if gender were an influence on productivity when organization variables on one hand and personal research styles and problem choice preferences on the other were included. Gender continued to be an important and necessary predictor in the limited data set studied here. I am open-minded as to whether with a considerably larger data set models of productivity will continue to require gender as a predictor. Gender is such a complex variable, a proxy for so many as yet undisclosed attitudes (and perhaps innate predispositions), that it may not be redundant in predictive models of productivity for the foreseeable future.

ENDNOTES TO CHAPTER SIX

[1] Partial analysis by discipline of the relationship between gender and competition both in current and in past projects yielded inconsistent results that suggested that the small sample size unduly influenced them. For example, when respondents were asked about trying to beat a rival for priority in the past there was a significant relationship between gender and competing for priority only within the exact sciences. However, in the current project, there was a strong relationship between gender and competing for priority among biologists and physicians. If possible, this issue should be explored with a larger data set in the future.

I also explored the relationship between professional autonomy and competition for priority. For this purpose I created a new variable of "ever competed for priority by combining variables measuring (a) whether competing for priority was a criterion in the past and (b) whether it was a criterion in the latest project. Although there was a difference between those in disciplines with weak paradigms and those in strong paradigms in this comparison, it was only marginally significant (Fisher's Exact Test, one-sided (P=.098). The weak statistical relationship could be partly because of the limited sample size.

[2] The multiple regression analysis reported here is based on Dr. Lawrence Lessner's replication of the author's own work. The principal difference between the analysis done by the author and the analysis by Lessner, a mathematical statistician and bio-statistician, was a consideration of the best *model* as opposed to simply trying to determine the best predictors.

[3] It was necessary to examine the residuals for this data set because there were several male respondents with very large values for productivity, and the possibility that these large values observations would influence the analysis was a matter of concern. There were 14 covariate patterns in the data and the Pearson and Deviance residual plots did not show any alarming patterns. Only one observation, obs=13, was larger than 2.0: his standardized Pearson residual=2.40. This was a male respondent with 620 for production. Actually there were three other male respondents with larger values of production that were not identified as extreme: PROD = 745, 747, 1795. When the analysis was rerun all the coefficients did change but only "Prefer Problems No One Else-B" was no longer Wald significant and no longer significant in the type three sum of squares. While this is a significant change that was due entirely to observation 13, observation 13 was not the largest value of PROD. Furthermore, the extreme value of 2.4 is unlikely but not that rare an occurrence. The obs. 13 may be unusual in this small sample but probably belongs in the same group.

All efforts to identify organization variables that could predict: "Want to contribute to publication of a technical paper in a new dynamic area"; "Participation in MDR allows me to associate with technically superior colleagues"; and "Prefer Problems No One Else is Working on" were unsuccessful. Neither was SEX a predictor of any of these research style and/or problem choice criteria variables in any logistic regression models tried.

[4]According to Valian [1998:265], men's work is cited more often. However, this is because men do more papers. She says, "Men increase their overall citation rate by publishing more often."

BIBLIOGRAPHY

Aczel, Amir D.
1996 *Fermat's Last Theorem* New York: Four Walls Eight Windows

Aldrich, H. and Pfeffer, J.
1976 "Environments of Organizations," *Annual Review of Sociology*, 2:79-105 cited in Shenhav (1985)

Allen, Clark Lee
1962 *Elementary Mathematics of Price Theory* Belmont, CA: Wadsworth Publishing Company, Inc.

Astin, H.S.
1969 *The Woman Doctorate in America* New York: Russell Sage Foundation in Cole, J. (1979)

Banfield, Edward C.
1967 *The Moral Basis of a Backward Society* New York: The Free Press paperbound (originally 1958)

Barber, Bernard
1952 *Science and the Social Order* Glencoe, IL: Free Press cited in Zuckerman (1978)

Barnes, B.
1974 *Scientific Knowledge and Sociological Theory*, London & Boston, MA: Routledge and Kegan Paul cited in Shenhav (1985)

Berelson, Bernard R, Paul F. Lazarsfeld, and William N. McPhee
1966 *Voting*: Chicago: The University of Chicago Press Phoenix (softcover) (originally 1954)

Biddle, Bruce J. and Edwin J. Thomas, ed
1966 *Role Theory: Concepts and Research* New York: John Wiley & Sons, Inc.

Black, Max (ed)
1976 *The Social Theories of Talcott Parsons* Carbondale, Ill.: Southern Illinois University Press Acturur paperbacks (orig. 1961)

Boulding, Kenneth E.
1963 *Conflict and Defense* New York: Harper Torch Books paperbound (orig. 1962 Harper & Row)

Blalock, Hubert M., Jr.
1969 *Theory Construction* Englewood Cliffs NJ: Prentice-Hall Inc.

Blalock, Hubert M., Jr.
1990 *Power and Conflict* Newbury Park: Sage Publications

Bok, Sissela
1982 *Secrets* New York Pantheon Books

Braybrooke, David and Charles E. Lindblum
1970 *A strategy of Decision* New York: The Free Press (orig. 1963).

Busch, Lawrence, William B. Lacy, and Carolyn Sachs
1983 "Perceived Criteria for Research Problem Choice in the Agricultural Sciences - A Research Note *Social Forces* 62:1, 190-200

Callon, Michel

1995 "Four Models for the Dynamics of Science" in Jasanoff, Markle, Petersen, and Pinch eds. *Handbook of Science and Technology Studies* pp. 29-63

Caplow, Theodore
1964 *Principles of Organization* New York: Harcourt Brace & World, Inc.

Caplow, Theodore
1968 *Two Against One* Englewood Cliffs, NJ: Prentice-Hall, Inc.

Carroll, John S. and Eric J. Johnson
1990 *Decision Research: A Field Guide* Applied Social Research Methods series: v.22 Newbury Park Cal. Sage Publ. Inc.

Chapman, William R
1986 *Delinquency Theory and Attachment to Peers*
Ph.D. diss. University at Albany

Chein, Isidor
1976 "An Introduction to Sampling" in Selltiz, Claire, Lawrence S. Wrightsman, and Stuart W. Cook (eds.) *Research Methods in Social Relations*, 3rd edition, New York etc.: Holt, Rinehart and Winston

Chubin, Daryl E. and Alan Ll Porter, Frederick A. Rossini, and Terry Connolly eds.
1968 *Interdisciplinary Analysis and Research*
Mt. Airy, MD: Lamond Publications Inc.

Clegg, Stewart R., Cynthia Hardy and Walter R. Nord eds.
1996 *Handbook of Organization Studies* London: Sage Publications Ltd.

Cohen, M.D., March, J.G. and Olsen, J.P.
1972 "A Garbage Can Model of Organizational Choice," *Administrative Science Quarterly* 17:1-25

Cole, Johnathan R.
1979 *Fair Science*
New York: The Free Press

Cole, Stephen
1992 *Making Science*
Cambridge, MA: Harvard University Press

Cole, Stephen and Thomas J. Phelan
1999 "The Scientific Productivity of Nations," *Minerva*:37:1-23

Collins, R.
1975 *Conflict Sociology*
New York: Academic Press

Collins, H.M.
1982 "Knowledge, Norms, and Rules in the Sociology of Science" *Social Studies of Science*, 12:299-308
cited in Shenhav (1985)

Converse, Jean M. and Stanley Presser
1986 *Survey Questions* Sage University Paper Series on Quantitative Applications in the Social Sciences, Series No. 07-063 Newbury Park: Sage Publications

Cook, Thomas D. and Donald T. Campbell
1979 *Quasi - Experimentation*
Boston: Houghton-Mifflin Co.

Coser, Lewis
1964 *The Functions of Social Conflict*
New York: The Free Press (orig. publ. 1956)

Coser, Lewis A. ed.
1965 *Georg Simmel*
Englewood Cliffs, N.J.: Prentice Hall, paperbound

Cozzens, Susan
1989 *Social Control and Multiple Discovery in Science: The Opiate Receptor Case* Albany, NY: State University of New York quoted in White, H.C. (1992)

Crane, Diana
1977 "Social Structure in a Group of Scientists: A Test of the "Invisible College" Hypothesis in Leinhardt, Samuel (ed.) pp 161-178 (originally ASR, 34, 335-352 (1969)
Darwin, C.

Dabaekere, Koenraad and Michael A. Rappa
1994 "Institutional Variations in Problem Choice and Persistence Among Scientists in an Emerging Field
Research Policy: 23 pp. 425-441

Dariwin, Charles
1958 *The Origin of the Species* Mentor, 6th ed. (originally 1859)

Denzin, Norman K. (ed.)
1970 *Sociological Methods*
Chicago: Aldine Publishing

DiMaggio, Paul J. and Walter W. Powell
1983 "The iron cage revisited: institutional isomorphism and collective rationality in organizational fields."
American Sociological Review, 48:147-160

Donovan, Arthur, Larry Laudan and Rachel Laudan, ed.
1992 *Scrutinizing Science*
Baltimore, MD: The Johns Hopkins University Press (orig. publ. 1988)

Duffy, John
1993 *Economics*
Lincoln, Neb.: Cliff Notes

Ekeh, Peter
1974 *Social Exchange Theory*:
London: Heinmann paperback

Elster, Jon, ed.
1986 *Rational Choice*
New York University Press

Emerson, R.
1962 "Power-Dependence Relations"
American Sociological Review, 27:21-32 cited in Shenhav (1985)

Evan, William M., ed.
1976 *Inter-organizational Relations*
Harmondsworth, Middlesex, England: Penguin Books Ltd.

Feldt, Barbara
1986 "The Faculty Cohart Study: School Medicine." Ann Arbor Mich.: Office of Affirmative Action cited in Fox 1993

Festinger, Leon
1962 *A Theory of Cognitive Dissonance*
Stanford, California: Stanford University Press (originally published 1957 Row, Peterson & Co.)

Fisher, Joel L.
n.d. "Personal Communication" (1996)

Fisher, Robert L.
1984 'A Political Milieu of Evaluation Research' Paper Read at Annual Meeting of Academy of Criminal Justice Sciences, Chicago, Ill.

Fisher, Robert L.
1993 "Proposal for a Study of Problem Choice in Multidisciplinary Research Environments"
Unpubl. Ms. Department of Sociology, Columbia University in the City of New York

Fisher, Robert L., William Hopkins, Audrey Ignatoff and Donald DesJarlais
1979 "Female Sex" and "Percent Females in Progam" as Predictors of Success in a High School Drug Abuse Prevention and Education Program"
Unpubl. Ms. NYS Division of Substance Abuse Services

Fox, Mary Frank
1992 "Gender, Environmental Milieu, and Productivity in Science" in Zuckerman, H. Cole, J.R., and Bruer, J.T. (Ed.) *The Outer Circle*, pp 188-204

Fox, Mary Frank
1995 "Women and Scientific Careers" in Jasanoff, Markle, Petgersen, Pinch eds. *Handbook of Science and Technology Studies* (full citation under Jasanoff) pp. 205-223

Fox, R.
1976 "Scientific Enterprise and the Patronage of Research in France 1800-1870" in Turner, G.L'E. (ed) *The Patronage of Science in the Nineteenth Century*, Leyden; Noordhoff International Publishing cited in Shenhav (1985)

Freeman, J. H. and Hannan, M.T.
1975 "Growth and Decline Process in Organizations" *American Sociological Review*, 49:571-580 in Shenhav (1985)

Freeman, John J. and Michael T. Hannan
1983 "Niche Width and the Dynamics of Organizational Populations" *American Journal of Sociology* 88:1116-1145

Frost, Peter J., Larry F. Moore, Meryl Reis Louis, Craig C. Lundberg and Joanne Martin, eds.
1991 *Reframing Organizational Culture*
Newbury Park: Sage Publications

Fujimura, J.H.
1987 "Constructing 'Do-able' Problems in Cancer Research: Articulating Alignment"
Social Studies of Science 17 257-293

Fujimura, J.H
1988 "The Molecular Biological Bandwagon in Cancer Research: Where Two Worlds Meet" *Social Problems* 35 261-283

Garfinkel, Harold
1967 *Studies in Ethnomethodology*
Englewood Cliffs, NJ: Prentice-Hall, Inc.

Gaston, J.
1973 *Originality and Competition in Science*
Chicago: University of Chicago Press, quoted in Shenhav (1985)

Giele, Janet Z.
1988 "Gender and Sex Roles: in Smelser, Neil J. (eds) *Handbook of Sociology*

Giere, Ronald N.
1988 *Explaining Science*
Chicago: University of Chicago Press, paperbound 1990

Gieryn, T.H.
1978 "Problem Retention and Problem Change in Science" in Gaston, J. (ed.) *Sociology of Science* San Francisco: Jussey-Bass Publishers

Gieryn, T.H.
1983 "Boundary-Work and the Demarcation of Science From Non-science *American Sociological Review*, 48:781-795

Gieryn, T.H.
1995 "Boundaries of Science" in Jasanoff, Markle, Petersen, and Pinch eds. *Handbook of Science and Technology Studies* pp. 393-443

Gleick, James
1987 *Chaos* New York: Viking Penguin

Gouldner, Alvin W.
1965 *Wildcat Strike* New York: Harper Torchbooks (paperbound) originally published 1954 by Harper & Row

Gouldner, Alvin W.
1970 *The Coming Crisis of Western Sociology* New York:Basic Books

Graham, Loren
1987 *Science Philosophy, and Human Behavior in the Soviet Union* New York: Columbia University Press

Greenberg, D.S.
1966 "Bootlegging: It Holds a Firm Place in Conduct of Research" *Science,* 153:848-849 in Shenhav (1985)

Hagstrom, W. O.
1986 "The Differentiation of Disciplines in Chubin, Daryl E., Alant, Porter, Frederick A. Rosini, and Terry Connolly eds. *Interdisciplinary Analysis and Research* pp. 47-52

Hardy, Melissa A
1993 *Regression with Dummy Variables*
Newbury Park CA: Sage Publications

Herrnstein, Richard J. and Charles Murray
1994 *The Bell Curve*
New York: Free Press Paperback

Hilgartner, Stephen
1995 "The Human Genome Project" in Jasanoff, Markle, Petersen, and Pinch eds. *Handbook of Science and Technology Studies* pp. 302-316

Jacobs, Bruce A.
1996 "Crack Dealers Apprehension Avoidance Techniques: A Case of Restrictive Deterrence"
Justice Quarterly 13:3 pp. 359-381

Jackson, Douglas N. and J. Philippe Rushton, eds.
1987 *Scientific Excellence*
Newbury Park: Sage Publications

Janis, Irving L. and Leon Mann
1977 *Decision Making*
New York: The Free Press

Jasanoff, Sheila, Gerald E. Markle, James C. Petersen and Trevor Pinch, eds.
1995 *Handbook of Science and Technology Studies*
Thousand Oaks: Sage Publications

Jones, A.J.
1980 *Game Theory: Mathematical Models of Conflict*
New York: John Wiley & Sons

Junker, Buford H.
1972 *Field Work: An Introduction to the Social Sciences* Chicago: University of Chicago Press (orig. publ. 1960)

Kadushin, Charles
1966 "The Friends and Supporters of Psychotherapy: on Social Circles in Urban Life" ASR 31 (Dec.) 786-802

Kadushin, Charles
1968 "Power, Influence and Social Circles: a New Methodology for Studying Opinion Makers" ASR 33 (October) 685-699

Kadushin, Charles
1976 "Networks and Circles in the Production of Culture" in Peterson, A.R.(ed) *The Production of Culture* Beverly Hills and London: Sage Publications

Kim, Jae-On and Charles W. Mueller
1978 *Factor Analysis*
Sage University Paper Series on Quantitative Applications in the Social Sciences, Series No. 07-014 Newbury Park: Sage Publications

Kingdon, John W.
1995 *Agendas, Alternatives, and Public Policies* (second ed.) New York: Harper Collins College Publishers (paperbound) (first ed. published 1984)

Knorr-Cetina, K.
1982 "Scientific Communities of Transepistemic Arenas of Research: A critique of Quasi-Economic Models of Science" *Social Studies of Science* 12(1) 101-130

Knorr-Cetina, K.
1995 "Laboratory Studies: The Cultural Approach to the Study of Science" in Jasanoff, Sheila and Others (eds) *Handbook of Science and Technology Studies*

Kriesberg, Louis
1973 *The Sociology of Social Conflicts*
Englewood Cliffs, NJ: Prentice-Hall, Inc.

Kuhn, Thomas S.
1970 *The Structure of Scientific Revolutions*
Chicago The University of Chicago Press [Second Ed. enlarged (orig. publ. 1962)]

Landsberger, Henry A.
1976 "Parsons' Theory of Organizations" pp, 214-249 in Max Black (ed.) *The Social Theories of Talcott Parsons* Carbondale and Edwardsville
Illinois Southern Illinois University Press paperbound

LaPorte, T.R. and J.L. Wood
1970 "Functional Contributions of Bootlegging and Entrepreneurship in Research Organizations"
Human Organization 29:273-287
cited in Shenhav (1985)

Latour, B. and S. Woolgar
1979 *Laboratory Life: The Social Construction of Scientific Facts*
Beverly Hills CA: Sage Publications

Lawrence, Paul R. and Jay W. Lorsch
1969 *Organization and Environment*
Homewood Il: Richard D. Irwin

Lazarsfeld, Paul F and Morris Rosenberg ed.
1995 *The Language of Social Research*
New York: The Free Press

Lazarsfeld, Paul F. and Neil W. Henry, ed.
1966 *Readings in Mathematical Social Science*
Cambridge, MA the M.I.T. Press

Lazarsfeld, Paul F., Jeffrey G. Reitz, and Ann K. Pasanella
1975 *An Introduction to Applied Sociology*
New York: Elsevier Scientific Publishing Company

Lefebvre, Henri
1969 *The Sociology of Marx* trans. Norbert Guterman (esp. Chap. 3 "Ideology and the Sociology of Knowledge")
New York: Vintage Books paperbound (orig. English Edition 1968 Random House)

Leinhardt, Samuel ed.
1977 *Social Networks*
New York: Academic Press, Inc.

Litwak, Eugene and Henry J. Meyer
1974 *School, Family, and Neighborhood: the Theory and Practice of School-Community Relations*
New York: Columbia University Press

Long, J.S. and McGinnis R.
1981 "Organizational Context and Scientific Productivity"
American Sociological Review 46:422-442

March James G.
1988 *Decisions and Organizations*
Oxford: Basil Blackwell Ltd.

Marx, Karl
1967 *Writings of the Young Marx on Philosophy and Society* (trans. and edited Easton, L.D. and K.H. Guddat)
Garden City, NY: Anchor Books paperbound

Marx, Karl
1964 *Selected Writings in Sociology and Social Philosophy* trans. T.B. Beltomore
New York: McGraw-Hill Paperback ed. (orig. C.A. Watts, Ltd. London, 1956)

McLeod, R.
1977 "The Social History of Science"
In Spiegel-Rosing, Ina and Derek J. de Sola Price (eds.) *Science, Technology and Society: A Cross-Disciplinary Perspective* Beverly Hills CA: Sage Publications

Merton, Robert K.
1964 *Social Theory and Social Structure*
London: The Free Press of Clencoe (Orig. publ. 1957)

Merton, Robert K.
1967 *On Theoretical Sociology*
Toronto, Canada: Collier-Macmillan Canada Ltd. (Free Press paperbound)

Merton, Robert K.
1973 *The Sociology of Science*
Chicago: The University of Chicago Press

Merton, Robert K., Leonard Broom, and Leonard S. Cottrell, Jr. (eds)
1965 *Sociology Today*
New York: Harper & Row (paperbound) originally Basic Books, Inc. 1959

Meyer, J.W. and M. Rowan
1977 "Institutional Organizations: Formal Structure as Myth and Ceremony" *American Journal of Sociology* 83:340-363

Miles, Matthew B. and A. Michael Huberman
1994 Qualitative Data Analysis, 2nd ed.
Thousand Oaks: Sage Publications

Miller, Susan, David J. Hickson, and David C. Wilson
1996 "Decision-Making in Organizations" in Clegg, Hardy, and Norel (eds)
Handbook of Organization Studies

Mishan, E.J.
1971 *Cost-Benefit Analysis*
New York: Praeger Publishers

Moulin, Herve
1982 *Game Theory for the Social Sciences*
New York: New York University Press

Mulkay, Michael
1991 *Sociology of Science*
Bloomington, IN: Indiana University Press

Nederhof, Anton J. and Arie Rip
n.d. "Research Decisions: Doing Your Own Thing? Influences on Biotechnologists' Choices for Fundamental and Application-Oriented Research (Unpublished manuscript, 1986?)

Nelkin, Dorothy
1987 *Selling Science*
New York: W.H. Freeman and Company

Nozick, Robert
1993 *The Nature of Rationality*
Princeton NJ: Princeton Unviersity Press
Ofshe, Lynne and Richard Ofshe

Nozick, Robert
1970 *Utility and Choice in Social Interaction*
Englewood Cliff, NJ: Prentice-Hall, Inc.

Parson, Keith (ed.)
2003 *The Science Wars*
Amherst, NY: Prometheus Books

Parsons, Talcott
1960 *Structure and Process in Modern Societies*
New York: The Free Press

Parsons, Talcott and Neil J. Smelser
1965 *Economy and Society*
New York: The Free Press paperbound (orig. publ. 1956)

Pelz, Donald C. and Frank M. Andrews
1966 *Scientists in Organizations*
New York: John Wiley and Sons

Pfeffer, J. and Salancik, G.R.
1978 *The External Control of Organizations*
New York: Harper and Row cited in Zeldenrust (1990)

Pirsig, Robert M .
1984 *Zen and the Art of Motorcycle Maintenance*
New York: Bantam Books paperback (orig. publ. 1974)

Price, D.J. de.S.
1961 *Science Since Babylon*
New Haven, CT: Yale University Press 1963
Little Science, Big Science New York: Columbia University Press in Shenhav (1985)

Price, Don K.
1962 *Government and Science*
New York: Oxford University Press
(orig. publ. 1954)

Ragin, Charles C. and Howard S. Becker, eds.
1992 *What Is A Case?*
Cambridge: Cambridge University Press

Renzetti, Claire M. and Raymond M.Lee (eds)
1993 *Research On Sensitive Topics*
Newbury Park, CA: Sage Publications

Riker, William H.
1962 *The Theory of Political Coalitions*
New Haven, CT: Yale University Press

Sackmann, Sonja A.
1991 *Cultural Knowledge in Organizations*
Newbury Park: Sage Publications

Scott, W. Richard
1995 *Institutions and Organizations*
Thousand Oaks, CA: Sage Publications

Schelling, Thomas C.
1963 *The Strategy of Conflict*
New York Oxford University Press (orig. publ. 1960)

Shenhav, Yehouda A.
1985 Environmental and Organizational Effects on Research Agenda: The Issue of Funding
Ph.D. diss. Stanford University

Siegel, Sidney
1956 *Nonparametric Statistics*
New York: McGraw Hill Book Co., Inc.

Simmel, G.
1950 *The Sociology of Georg Simmel*
Wolff, K.H.(ed) New York: The Free Press: 162-169

Simmel, G.
1966 *Conflict and The Web of Group-Affiliations* translated by Kurt H. Wolff and Reinhard Bendix
New York: The Free Press [(orig. publ. 1955)]

Smelser, Neil J., ed.
1988 *Handbook of Sociology*
Newbury Park: Sage Publications

Spiegel-Rasing, I. and D. de S. Price (eds.)
1977 *Science, Technology and Society - A Cross Disciplinary Perspective* CA: Sage Publications Inc.

Spykman, Nicholas J.
1966 *The Social Theory of Georg Simmel*
New York: Atherton Press

Stigler, George J.
1966 *The Theory of Price*
New York: The Macmillan Company

Strauss, Anselm and Juliet Corbin
1990 *Basics of Qualitative Research*
Newbury Park: The Sage Publications

Sulloway, Frank J.
1996 *Born to Rebel*
New York: Pantheon Books

Tatsuoka, Maurice
1971 *Multivariate Analysis*
New York: John Wiley & Sons, Inc.

Thagard, Paul
1992 *Conceptual Revolutions*
Princeton NJ: Princeton University Press (paperback 1993)

Thompson, James D.
1967 *Organization in Action*
New York: McGraw-Hill Books

Useem, M.
1976a "State Production of Social Knowledge: Patterns in Government Financing of Academic Social Research"
American Sociological Review 41:613-629

Useem, M
1976b "Government Influence on the Social Science Paradigm" *The Sociological Quarterly* 17:146-161

Velho Lea
1990 "Sources of Influence on Problem Choice in Brazilian University Agricultural Science" *Social Studies of Science* 20:503-17

cman, Judy
1995 "Feminist Theories of Technology" in Jasanoff, Markle, Petersen, and Pinch eds. *Handbook of Science and Technology Studies* pp. 189-204

Watson, James
1968 *The Double Helix*
New York: Atheneum

Westfall, R.E.
1977 *The Construction of Modern Science*
Cambridge: Cambridge University Press in Shenhav (1985)

White, Harrison C.
1992 *Identity and Control*
Princeton, NJ: Princeton University Press

Wicklund, Robert A. and Jack W. Brehm
1976 Perspectives on Cognitive Dissonance
Hillsdale, NJ: Lawrence Erlbaum Associates, Inc.

Wolff, Kurt H. trans and ed.
1964 *The Sociology of Georg Simmel*
New York: The Free Press (orig. publ. 1950)

Wuthnow, R.
1980 "The World-Economy and the Institutionalization of Science in Seventeenth-Century Europe" in Bergesen, A (ed.) *Studies of the Modern World System*
New York: Academic Press

Xie, Yu and Shauman, Kimberly A.
1998 "Sex Differences in Research Productivity: New Evidence About An Old Puzzle," *American Sociological Review*, vol. 63 (Decemeber: 847-870)

Zeldenrust, Sjerp
1990 *Ambiguity, Choice and Control in Research*
Ph.D. diss. Universiteit van Amsterdam, Netherlands

Ziman, J.M.
1983 "The Collectivization of Science" Proceedings of the Royal Society (London)
B 219 (1983) 1-19

Ziman, J.M.
1987 "The Problem of 'Problem Choice'"
Minerva XXV, #1-2, 92-106

Zuckerman, Harriet
1977 *Scientific Elite*
New York: The Free Press

Zuckerman, Harriet
1978 "Theory Choice and Problem Choice in Science" In: J. Gaston (ed.), *The Sociology of Science - Problems, Approaches and Research*
San Francisco: Jossey-Bass

Zuckerman, Harriet
1988 "The Sociology of Science," in N. Smelser (ed.) *Handbook of Sociology,* pp. 511-574

Zuckerman, Harriet
1991 "The Careers of Men and Women Scientists," in Zuckerman, H., Cole J.R., and Bruer, J.T. (ed.) *The Outer Circle* pp: 27-56

Zuckerman, Harriet, Jonathan R. Cole and John T. Bruer eds.
1991 *The Outer Circle*
New York: W.W. Norton & Co. (reprinted paperbound in 1993 by Yale University Press, New Haven, CT

INDEX

A

B

C

D

E

F

G

L

M

N

O

P

Q

R

S

T

U

V

W

X

Z